U.S. Department of Transportation
National Highway Traffic Safety Administration

DOT HS 811 681

November 2012

Evaluation of the Effectiveness Of TPMS in Proper Tire Pressure Maintenance

DISCLAIMER

This publication is distributed by the U.S. Department of Transportation, National Highway Traffic Safety Administration, in the interest of information exchange. The opinions, findings, and conclusions expressed in this publication are those of the authors and not necessarily those of the Department of Transportation or the National Highway Traffic Safety Administration. The United States Government assumes no liability for its contents or use thereof. If trade names, manufacturers' names, or specific products are mentioned, it is because they are considered essential to the object of the publication and should not be construed as an endorsement. The United States Government does not endorse products or manufacturers.

Suggested APA Format Citation:
Sivinski, R. (2012, November). *Evaluation of the effectiveness of TPMS in proper tire pressure maintenance.* (Report No. DOT HS 811 681). Washington, DC: National Highway Traffic Safety Administration.

Technical Report Documentation Page

1. Report No. DOT HS 811 681	2. Government Accession No.	3. Recipient's Catalog No.
4. Title and Subtitle Evaluation of the Effectiveness of TPMS in Proper Tire Pressure Maintenance		5. Report Date November 2012
		6. Performing Organization Code
7. Author Robert Sivinski		8. Performing Organization Report No.
9. Performing Organization Name and Address Evaluation Division; National Center for Statistics and Analysis National Highway Traffic Safety Administration Washington, DC 20590		10. Work Unit No. (TRAIS)
		11. Contract or Grant No.
12. Sponsoring Agency Name and Address National Highway Traffic Safety Administration 1200 New Jersey Avenue SE. Washington, DC 20590		13. Type of Report and Period Covered NHTSA Technical Report
		14. Sponsoring Agency Code
15. Supplementary Notes		

16. Abstract

This report is an analysis of the data collected through the Tire Pressure Monitoring System-Special Study as it pertains to the effectiveness of TPMS in promoting proper tire inflation. The study was conducted in 2011, using a nationally representative sampling structure, based on the primary sampling units (PSUs) of the National Automotive Sampling System. NASS personnel collected 6,103 complete vehicle observations including tire pressure of all four tires. This survey found that 23.1 percent of the MY 2004-2007 vehicles without TPMS had at least one severely underinflated tire as defined by FMVSS No. 138 (25% or more below the vehicle manufacturer's recommended cold tire pressure), but only 11.8 percent of the MY 2004-2007 vehicles equipped with TPMS had a severely underinflated tire. Based on these results, the presence of TPMS on a vehicle of model years 2004 to 2007 is estimated to result in a 55.6-percent reduction in the likelihood that the vehicle will have one or more severely underinflated tires as defined by FMVSS No. 138. It is also estimated to result in a 30.7-percent reduction in the likelihood that the vehicle will have one or more tires that are overinflated by 25 percent or more above the manufacturer's recommended cold tire pressure. During the first eight years of operation TPMS is estimated to save a typical passenger car 9.32 gallons of fuel and a typical LTV 27.89 gallons of fuel. During 2011 TPMS is estimated to have saved $511 million across the vehicle fleet through reduced fuel consumption. NHTSA plans to conduct further research to determine the effect of TPMS on the incidence of tire-related crashes and injuries.

17. Key Words NHTSA; TPMS; Tire Pressure Monitoring System; inflation; FMVSS 138; effectiveness	18. Distribution Statement Document is available to the public from the National Technical Information Service www.ntis.gov		
19. Security Classif. (Of this report) Unclassified	20. Security Classif. (Of this page) Unclassified	21. No. of Pages 58	22. Price

Form DOT F 1700.7 (8-72)

Table of Contents

Executive Summary iv

1. **Introduction**
   1.1: Risks of Over/Underinflation 1
   1.2: Tire Pressure Monitoring Systems 1
   1.3: Past Research 4
   1.4: Goals of the Evaluation 4

2. **Survey Methods**
   2.1: Survey Design 5
   2.2: Site Cooperation 6
   2.3: Survey Participants 7
   2.4: Data Collection Staff 7
   2.5: Data Collection Schedule 8
   2.6: Data Collection Equipment 8
   2.7: Forms and Variables 9
   2.8: Questionnaire Development and Training 10
   2.9: Data Entry and Quality Control 11
   2.10: Weighting and Estimation 11
   2.11: Data Corrections and Adjustments 12

3. **Underinflation**
   3.1: Summary 13
   3.2: Population Estimates 13
   3.3: TPMS Effectiveness 15
   3.4: Effectiveness by Display and System Type 18
   3.5: Effectiveness by Vehicle Type 20

4. **Overinflation**
   4.1: Summary 22
   4.2: Population Estimates 22
   4.3: TPMS Effectiveness 24
   4.4: Effectiveness by Display and System Type 26
   4.5: Effectiveness by Vehicle Type 27

5. **Fuel Economy**
    5.1: Summary 28
    5.2: Average Underinflation 28
    5.3: Effect of TPMS on Fuel Economy 30

6. **Discussion**
    6.1: Summary of results 36
    6.2: Possible Sources of Bias 37
    6.3: Limitations 38
    6.4: Future Research 39

7. **Appendix**
    7.1: Weighting A1
    7.2: Survey Forms A4

Executive Summary

This report presents an analysis of the data collected through the Tire Pressure Monitoring System-Special Study (TPMS-SS) as it pertains to the effectiveness of TPMS in promoting proper tire inflation. The TPMS-SS was conducted in 2011, using a nationally representative sampling structure based on the primary sampling units (PSUs) of the National Automotive Sampling System. NASS personnel collected 6,103 complete vehicle observations (4,391 of which were equipped with TPMS) including tire pressure and temperature of all four tires. Only vehicles in the model year range 2004-2011 were surveyed.

Proper tire inflation is important for several reasons. Underinflated tires experience a greater amount of sidewall flexion than properly inflated tires, resulting in decreased fuel economy, sluggish handling, longer stopping distances, increased stress to tire components, and heat buildup that can lead to catastrophic failure of the tire, such as cracking, component separation, or blowout. These catastrophic failures can cause loss of vehicle control and may result in a crash. Overinflated tires may be more easily damaged by potholes or debris. Severe overinflation may increase stopping distance due to reduced area of road contact and non-optimal traction, and may also contribute to vehicle instability. As with underinflation, overinflation may result in uneven tread wear that reduces the useful life of the tire. This report does not attempt to directly measure the relationship between TPMS and vehicle safety; rather it measures the relationship between TPMS and proper tire inflation. In order to estimate the effect that TPMS has on crash avoidance and mitigation, future analyses are planned that will use real-world crash data.

This survey found that 12.4 percent of all passenger vehicles in the US of model years 2004-2011 have at least one tire that is severely underinflated as defined by FMVSS No. 138 (25% or more below the vehicle manufacturer's recommended cold tire pressure). The survey also found that 23.1 percent of the MY 2004-2007 vehicles without TPMS had at least one severely underinflated tire, but only 11.8 percent of the MY 2004-2007 vehicles equipped with TPMS – and only 5.7 percent of the more recent, MY 2008-2011 vehicles equipped with TPMS – had a severely underinflated tire. Based on the data from model year 2004-2007 vehicles (the range of collected model years that contained both vehicles with and without TPMS), TPMS was estimated to result in a statistically significant 55.6-percent reduction in the likelihood that a vehicle will have one or more severely underinflated tires.

TPMS is also estimated to result in a 30.7-percent reduction in the likelihood of severe overinflation (25% or more above the manufacturer's recommended cold tire pressure) for model year 2004-2007 vehicles. This effect was present in vehicles with TPMS systems that do not alert the driver to overinflation, and it is unclear what causes the association between TPMS and reduced overinflation. Although the specific effects of overinflation are not well documented, it is reasonable to assume that it results in more rapid tire wear and, possibly, reduced vehicle stability.

While the primary goal of TPMS is to reduce underinflation in order to make vehicles safer to operate, a further benefit of reduced underinflation is improved fuel economy. By combining estimates of reduced underinflation due to TPMS with estimates of increases in fuel economy resulting from increases in tire pressure, it's possible to estimate the amount of fuel that TPMS will save an average vehicle during a given period of time. During the first eight years of operation, TPMS is estimated to save a typical passenger car 9.32 gallons of fuel and a typical LTV 27.89 gallons of fuel. During 2011 TPMS is estimated to have saved $511 million across the vehicle fleet through reduced fuel consumption. This estimate does not include any additional savings that may result from extended tire life or any crash-avoidance benefits.

Future analyses of TPMS effectiveness will be conducted using crash data to directly measure the relationship between TPMS and tire-related crashes. Also, future analyses are planned to analyze the interview data collected through the TPMS-SS. This interview data includes information about driver behavior as it pertains to tire care and maintenance.

1: Introduction

1.1: Risks of Under/Overinflation

Proper tire inflation is important for several reasons. According to the Rubber Manufacturers' Association, underinflation is the most common cause of tire failure.[1] Underinflated tires experience a greater amount of sidewall flexion than properly inflated tires, resulting in decreased fuel economy, sluggish handling, longer stopping distances, increased stress to tire components, and heat buildup that can lead to catastrophic failure of the tire, such as cracking, component separation, or blowout.[2] These catastrophic failures can cause loss of vehicle control and may result in a crash. Overinflated tires may be more easily damaged by potholes or debris. Severe overinflation may increase stopping distance due to reduced area of road contact and non-optimal traction, and may also contribute to vehicle instability. As with underinflation, overinflation may result in uneven tread wear that reduces the useful life of the tire.

This report evaluates the effectiveness of tire pressure monitoring systems (TPMS) in reducing the frequency of severely underinflated tires by using survey data acquired through the Tire Pressure Monitoring System – Special Study (TPMS-SS), specifically designed for this purpose. (Future analyses are planned to estimate the effect that TPMS has on crash avoidance and mitigation, based on real-world crash data.) Statistical survey sampling techniques were used to generate a nationally representative sample of passenger vehicles of model years 2004-2011. The TPMS-SS collected data on 6,503 vehicles at gas stations throughout the United States from August 2010 through April 2011. Over 90 percent of these observations (n = 6,103) were complete enough to include in analyses of TPMS effectiveness. The basic method of analysis was to compare under- and overinflation rates observed in vehicles with and without TPMS. The magnitude of the difference in these rates formed the estimates of TPMS effectiveness.

1.2: Tire Pressure Monitoring Systems

Section 13 of the Transportation Recall Enhancement, Accountability, and Documentation (TREAD) Act, which Congress passed on November 1, 2000, directed NHTSA to conduct rulemaking actions to revise and update the Federal Motor Vehicle Safety Standards (FMVSS) for tires, to improve labeling on tires, and to require a system in new motor vehicles that warns the operator when a tire is significantly underinflated. In response, in 2005 NHTSA published the final rule for FMVSS No. 138 which requires that passenger cars, multipurpose passenger vehicles (MPVs), and trucks and buses with a gross vehicle weight rating (GVWR) of 10,000 pounds or less be equipped with a TPMS that is capable of detecting 25 percent underinflation in any combination of tires. The rule specified a phase-in schedule from 2005

---

[1] Rubber Manufacturers Association. "Tire and Auto Safety." [Online] Available at http://www.rma.org/tiresafety/auto_safety_facts.html, March 2001.
[2] Final Regulatory Impact Analysis, Tire Pressure Monitoring System, FMVSS No. 138, March 2005, NHTSA, Docket No. 2005-20586-2.

through 2007, with all vehicles manufactured on or after September 1, 2007, required to be equipped with TPMS.

These systems must alert the driver to low tire pressure (pressure 25% or more below the manufacturer's recommended cold tire inflation pressure, which is typically located on the door jamb label) through a dash displayed warning light. The display must activate within 20 minutes of underinflated travel at speeds of 50-100 km/hr and must remain illuminated until the underinflation is remedied. The system must also have a malfunction lamp in addition to a low pressure warning lamp that alerts the driver if the vehicle's TPMS is not functioning properly. Additional details of the requirements of FMVSS No. 138 can be found in the Code of Federal Regulations.[3]

Two distinct TPMS technologies have been installed in production vehicles at various times since 2000:

Direct TPMS: Also known as pressure sensor based (PSB), direct systems use a physical pressure sensor inside each tire that sends information to a central processing unit in the vehicle. Some direct systems exceed the minimum requirements of the standard and display the pressure of each tire on the dash allowing the driver to diagnose overinflation as well as underinflation. The system's sensors are most often located on the interior end of a tire's valve stems. Therefore, pressure information must be transmitted to the vehicle via a battery-powered radio frequency transmitter. Unlike indirect systems that do not have sensors in each wheel, direct systems may require additional maintenance and repair when pressure sensors malfunction due to harsh environmental conditions, depleted batteries or if they are removed, replaced or otherwise affected during tire repair or replacement.

Indirect TPMS: Also known as wheel speed based (WSB), indirect systems use individual wheel rotation speeds gathered from the anti-lock brake system (ABS) wheel speed sensors to detect relative underinflation. The operating principle is that if a single wheel has a faster rotational speed than other tires, then its radius or rolling circumference must be smaller and therefore the tire may be underinflated. Because indirect systems produced before FMVSS No.138 were only capable of diagnosing underinflation in relation to the other tires on the vehicle, these systems experienced difficulty identifying underinflated tires if several tires were simultaneously underinflated. FMVSS No. 138 required that monitoring systems must be capable of detecting underinflation in 'one or more of the vehicles' tires, up to a total of four tires', and as a result most manufacturers began exclusively installing direct systems after publication of the final rule. Fewer than three percent of the vehicles observed by the TPMS Special Study were equipped with indirect systems.

Recently some manufacturers have begun to produce indirect systems that integrate other types of data from the ABS and electronic stability control (ESC), along with information from other evolving technology sensors, into the diagnostic model. At the time of publication there are two vehicle manufacturers, Volkswagen and Audi, which utilize indirect systems in production vehicles. The first vehicle with an indirect system that met the requirements of FMVSS No. 138 was the model year 2009 Audi A6. In 2010 and 2011 both Audi and Volkswagen began to include indirect systems on several models including the Audi A3, A4, A5, A6, A8, Q5, and Q7and Volkswagen's Golf, GTI, and Jetta. Post-

---

[3] *Code of Federal Regulations,* Title 49, Part 571.138

FMVSS No.138 indirect systems may have many practical advantages over direct systems in terms of long-term maintenance because there is no physical sensor installed on the wheel and the systems do not require independent batteries. Although this survey did not collect sufficient data from these post-FMVSS No. 138 indirect systems to evaluate their effectiveness, their performance may be similar to the direct systems evaluated in this report (because they also comply with the requirements of FMVSS No. 138).

Display Types: To meet the requirements of FMVSS No.138 a vehicle is required to have a warning lamp that alerts a driver to severe underinflation as well as a malfunction lamp that indicates when the TPMS is not functioning properly. In addition, some vehicles are equipped with tire specific warning lamps that tell the driver which tire is triggering the underinflation warning. Other vehicles have tire specific inflation pressure readouts that are capable of reporting the current pressure of each tire at any time in addition to an underinflation warning lamp. These systems may alert drivers to less severe underinflation that has not yet reached the 25% threshold needed to activate the warning lamp. They may also prevent overinflation, which is not identified by other types of TPMS displays. Pressure displays are only possible with direct TPMS. These display types are evaluated separately in Section 3.4 (underinflation) and Section 4.4 (overinflation) of this report to explore any possible differences in effectiveness.

Market Share: Figure 1 below shows an estimate of the percent of new vehicle models sold with TPMS for model years 2000-2008 (TPMS became mandatory for all new vehicles manufactured on or after September 1, 2007). These estimates are based on Ward's Automotive Yearbook data for 98 popular vehicle models.

Figure 1: Estimated Percent of Vehicle Models Sold With TPMS

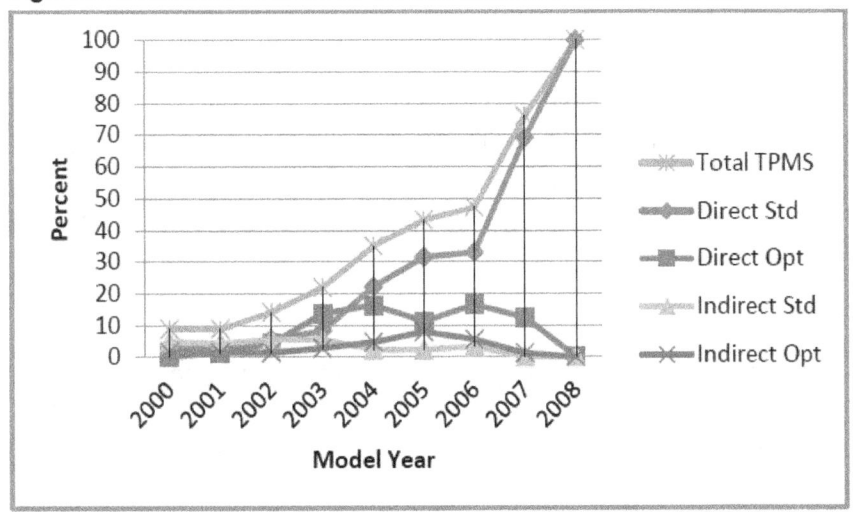

The steep rise in overall TPMS seen in 2007 is a result of the phase-in schedule specified by FMVSS No. 138, which required that at least 20 percent of all model year 2006 vehicles and 70 percent of all model year 2007 vehicles be equipped with TPMS. The graph also shows that indirect systems were not very common before the standard, and that none of the included vehicle models in model years 2006-2008 were equipped with an indirect system. (As mentioned above, new designs of indirect systems compliant with FMVSS 138 began to appear in model year 2009.)

1.3: Past Research

Tire Pressure Special Study (2001)[4] – This study used the 24 primary sampling units (PSUs) located throughout the United States established by the National Automotive Sampling System – Crashworthiness Data System (NASS-CDS). Within these PSUs, drivers were sampled for observation at gas stations from February 1, 2001, to February 14, 2001. According to the study, approximately 10 percent of the vehicles on the road in the United States were equipped with TPMS. A total of 11,530 vehicles were observed and the distribution of observed vehicle types (SUVs, vans, passenger cars, etc.) was consistent with national estimates of vehicle registration. There were no restrictions on the model years of included vehicles. The primary finding of this study was that 27 percent of passenger cars had at least one tire that was underinflated by at least eight psi (which would correspond to 25% underinflation for a tire with a common recommended pressure of 32 psi).

Tire Pressure Monitoring System Study (2009)[5] – This study also used the NASS-CDS infrastructure and was specifically designed to analyze the effectiveness of TPMS. This study collected data from a limited sample of 2,316 vehicles (data collection was halted before the planned sample size of 12,000 observations could be realized) from model years 1997 through 2003. The primary findings of this study were that most of the surveyed vehicles with correct tire pressure were equipped with TPMS. This study also found that 8.4 percent of observed vehicles with TPMS had at least one tire that was more than 25-percent underinflated, while 9.9 percent of observed vehicles without TPMS had at least one tire that exceeded the same level of underinflation. Most of the TPMS in this study (83%) were indirect systems.

1.4: Goals of the Evaluation

The primary goal of this study is to evaluate the effectiveness of TPMS in reducing underinflation in the on-road fleet of passenger vehicles, and this evaluation will be the focus of this report. A secondary goal of the study is to evaluate the effects of TPMS on overinflation. Although many systems are only equipped with a warning light that activates when a tire is underinflated, some systems offer real-time tire pressure displayed on the dash. These systems may have an effect on the rate of overinflation observed in vehicles equipped with them.

The third goal of this study is to estimate the observed improvement, if any, in fuel economy resulting from the presence of TPMS. Fuel savings resulting from increased tire pressure for vehicles with TPMS can at least partially defray the initial cost of the system.

---

[4] Bondy, N., Thiriez, K. (2001), *Tire Pressure Special Study Vehicle Observation Data.* (Report No. DOT HS 809 317), Washington, DC; National Traffic Safety Administration.
[5] Singh et al. (2009), *Tire Pressure Maintenance – A Statistical Investigation.* (Report No. DOT HS 811 086), Washington, DC; National Traffic Safety Administration.

The TPMS-SS survey also included driver interview items to measure driver knowledge about the importance of proper tire inflation and to identify the methods by which drivers are informed about issues pertaining to tire pressure maintenance. Subsequent analyses may be conducted to explore this interview data. This information could then inform data-driven and targeted behavioral programs that aim to promote proper tire inflation.

2: Methods

2.1: Survey Design

The NASS-CDS[6] infrastructure was used for data collection. This infrastructure offers several advantages and efficiencies for a nationally representative survey. The primary sampling units (PSUs) have already been selected and their weights have been computed. There are full-time data collectors at each PSU who are familiar with the area and are highly trained in interview techniques as well as data collection, data entry, and quality assurance. These factors allowed for a larger sample size to be collected within a set budget. The proposed survey design and the data collection forms were submitted to the Office of Management and Budget for approval, the approval number is 2127-0626.

Each of the 24 NASS-CDS PSUs were included in the study. Within each of the PSUs, ZIP codes were randomly selected for observation with equal probability of selection. Within each selected ZIP code, gas stations were first screened, based on inclusion criteria such as traffic flow, location, number of pumps, and presence of an attached convenience store in order to ensure that adequate data could be collected. Then, stations were selected from those meeting the inclusion criteria (this stage of selection was based on traffic flow and researcher convenience). Usually, two gas stations were selected per ZIP code and seven ZIP codes per PSU for a total of 14 gas stations within each of the 24 PSUs. Once a gas station was selected, a data-collection team consisting of an observer and an interviewer spent one day (eight hours) sampling vehicles that entered the station. Only vehicles of model year 2004 or newer were approached for inclusion due to the rarity of TPMS in older vehicles and to ensure that the number of sampled vehicles equipped with TPMS would be sufficient for statistical analysis.

In total, data from 6,503 passenger vehicles from model years 2004-2011 were collected. To be included in effectiveness estimates the minimum completeness requirements were measurements of pressure and temperature of all four tires, observed presence of TPMS system and type as well as vehicle make, model, and model year (used to confirm observed TPMS information). A total of 6,103 vehicles met these completeness requirements and were included in the following analyses of effectiveness.

As with any survey, the accuracy and reliability of the estimates derived in this report depend largely on the amount of data collected and the magnitude and variance of the sampling weights imposed by the sample design. Even with the large overall sample, if there are very few observations in any of the

---

[6] Additional information about the NASS-CDS PSUs can be found in Appendix "F" of the 2010 NASS-CDS Analytical User's Manual, which is located at the following Internet address: http://www-nrd.nhtsa.dot.gov/Pubs/NASS10.pdf.

subpopulations analyzed then the estimates derived may be unstable, non-significant, or otherwise misleading. The following tables give the unweighted counts of observations. The counts are separated into model year groups: 2004-2007, 2008-2011, and overall total. These model year groups are important because some of the evaluations in this report only use data from model years 2004-2007 in order to minimize an important possible source of bias due to vehicle age.

Table 1: Counts of Observations

Overall	MY 2004-2007	MY 2008-2011	Total
All Vehicles	3,297	2,806	6,103
Vehicle Type			
Passenger Car	1,565	1,643	3,208
Light Truck/van	1,732	1,163	2,895
TPMS			
With TPMS	1,585	2,806	4,391
Without TPMS	1,712	0	1,712
TPMS Type			
Indirect	209	0	209
Direct	1,376	2,806	4,182
TPMS Display Type			
Warning Lamp Only	1,248	2,031	3,279
Tire Specific Lamp	135	225	360
Tire Specific PSI	202	495	697
Unknown	0	55	55
Improper Inflation			
≥ 25% Underinflation	534	150	684
≥ 25% Overinflation	412	265	677

These counts are unweighted; however, all of the effectiveness estimates in the report were derived using weighted data.

2.2: Site Cooperation

Within each of the 24 NASS-CDS PSUs data were collected from August 2010 to April 2011. Cooperation of the owners of the gas stations at the different data collection sites was established prior to the commencement of the study mostly via in-person visits by NASS researchers. In addition, a letter explaining the scope of the survey was provided to the managers of the participating gas stations on the day of data collection. If the researchers were unable to collect data at a pre-authorized site due to unexpected circumstances, the next gas station listed for that ZIP code was substituted as an alternate site. At the conclusion of the data collection for the day, managers of the gas station were thanked and given a copy of the informational hand-outs that had been provided to the participants. Later, a thank-you letter was sent to the managers of the gas stations that participated in the study.

2.3: Survey Participants

The study's focus was on vehicles and their drivers, although some information was also collected on other passengers in the vehicles. Both the Tire Pressure Interview Form and the Refueling Interview Form (the survey forms are included in the appendix of this report) were used to screen and confirm that the vehicle's model year was 2004 or later. If the vehicle was not, the driver was thanked and the interview was ended.

To be included in the study a vehicle had to fall within one of four body types (i.e., passenger cars, utility vehicles, vans, and pickup trucks), the vehicle must have had a Gross Vehicle Weight Rating of less than 10,000 pounds, and the vehicle's model year must be 2004 or later. Vehicles without VINS or those with dual wheels on an axle were excluded as well as vehicles which did not enter the gas station to refuel. Some factors that did not affect whether a vehicle was selected for inclusion in the study were whether the driver: 1) drove a vehicle with TPMS, 2) was the vehicle's primary driver, or 3) was the person responsible for the vehicle's maintenance. On the other hand, only drivers of vehicles who were aware that their vehicle had TPMS were asked to answer the questions in the Supplemental Form.

If the researcher judged a vehicle to meet the inclusion criteria, the vehicle was approached after it had stopped at the gas pump. The researcher would give the driver a letter of introduction and ask the driver to participate in the study. Once a driver agreed to participate, one researcher interviewed the driver, recording information on the interview forms, while the second researcher inspected the vehicle and recorded data on the Vehicle Inspection Form and the Tire Inspection Form. Spanish language data collection forms were available, in case the interview could not be conducted in English. At some site, Spanish speaking interviewers were also available. At the conclusion of each interview, the participant was given a Courtesy Card. This card contained contact information (i.e., a TPMS study email address, DOT hotline telephone number) so that the drivers had someone to contact in case they had any further questions once they left the study site. Also included on the Courtesy Card were the air pressure measured on each tire, the vehicle manufacturer's recommended cold tire pressure, and several tire safety tips. The participants were also given several brochures with additional information about tire safety.

2.4: Data Collection Staff

Field data collection was conducted by two-person teams composed of trained data collectors. The materials employed included the interview forms, hand-outs for the participants, large sign(s) with information about the survey, special equipment (e.g., depth indicators, pyrometers, analogue air pressure gauges), DOT identification badges, the procedures manual, and miscellaneous items needed to obtain and record data such as clip boards, watches, and digital cameras. In general, one member of the team was responsible for collecting the interview data while the other was responsible for collecting vehicle data such as tire pressure.

2.5: Data Collection Schedule

Data were collected during the day, normally between the hours of 8 a.m. and 5 p.m., but some data were collected as early as 7 a.m. or as late as 6 p.m. The majority of the day was spent collecting data via interviews with the drivers and inspections of the vehicles; however, during each day there was a 45-minute lunch break as well as five 15-minute breaks spaced evenly throughout the day, during which time the researchers collected observational data on the vehicles that entered the gas station in order to estimate total traffic at the station for that day.

At the start of the day, the team would set up a "command center" at the gas station where extra survey forms and materials were kept. One or two 3' by 5' signs would be prominently displayed so drivers would be alerted to the fact that a tire pressure study was being conducted that day at the gas station.

2.6: Data Collection Equipment

Special Equipment used for data collection included an air pressure gauge to measure tire pressure, a tread depth indicator to measure tread depth, and a pyrometer to measure tire sidewall temperature and ambient air temperature.

The air pressure gauges were tested when received by the Zone Centers. In addition, they were tested at least 2 days prior to data collection and again at the beginning of each day's data collection. The test was conducted using the two air pressure gauges assigned to each team on the same tire of a vehicle. If the pressures were not within a 1 psi tolerance, they were to notify their Zone Center for immediate replacement. If the researcher could not determine which gauge was inaccurate, both gauges were to be replaced. If replacement gauges could not be in time for a scheduled day of data collection, data collection was to be postponed for that day and rescheduled for another day in the same week.

The pyrometers used in the study were checked against each other prior to each data collection day. If the pyrometers did not measure the same ambient air temperature (within a tolerance of one degree), the researchers noted the discrepancy on the Daily Site Form and notified the zone center for further direction.

No problems were noted with any of the equipment used in the study.

2.7: Forms and Variables

Data collected during the TPSS-SS included information on the sites at which data were collected, the vehicles that stopped to refuel at these sites, and the drivers and passengers of these vehicles. These data were recorded on seven data collection forms (see appendix) and were collected via observation, inspection and interview.

Observational Data

The observational data was collected on two forms: the Daily Site Form (Tallies & Inspections) and the Daily Site Form (Refueling). Every two hours, beginning at the initial time of data collection for the day, the data collectors observed, for a period of 15 minutes, the number of 2004-2011 passenger cars and light trucks that entered the gas station to obtain gasoline.

One researcher would tally, by body-type category (i.e., passenger cars, utility vehicles, vans, and pickup trucks) and time of day, the number of vehicles entering the gas station to obtain gasoline. In addition, the area around the gas station was characterized by the observer as being urban, suburban or rural in nature. This observational data was recorded in the Daily Site Form (Tallies & Inspections) and was used to explore non-response bias at the site and vehicle levels.

The other researcher would log the time it took for vehicles to refuel, by body type and time of day. In addition, the observer recorded: 1) the other activities in which the drivers engaged, such as going into the gas station store or putting oil into the vehicle, 2) characteristics of that particular station, such as type of payment used and additional services offered at this location (e.g., car wash, auto repair), and 3) whether the observation period ended before a specific driver had finished refueling. This observational data was recorded in the Daily Site Form (Refueling).

Inspection Data

The inspection data were collected on two inspection forms: the Vehicle Inspection Form and the Tire Inspection Form. Both forms were completed by one NASS researcher while another researcher conducted an in-person interview with the driver.

The Vehicle Inspection Form was used to obtain information about the vehicle (e.g., make, model, and model year) and the type of display (i.e., warning light only, tire specific warning icons, or tire specific display of psi) that was used to provide TPMS information to the driver. In addition, the tire-specific psi values seen in the TPMS displays, as well as information from the vehicle placard (i.e., recommended tire size, recommended air pressure, and the Gross Vehicle Weight Rating) were recorded on these forms.

The Tire Inspection Form was used to obtain information about all four of the vehicle's tires, including specifics about the tires (i.e., manufacturer, model, size, and recommended maximum pressure) and details about the condition of the tires at the time of the observation (i.e., measured temperature, pressure, and tread depth). In addition, the ambient air temperature and general weather conditions were recorded on the form.

Interview Data

Interview data were collected on three interview forms: the Tire Pressure Interview Form, the Refueling Interview form, and the Supplemental Form. Drivers of eligible vehicles were asked the questions on the interview forms, while only drivers of vehicles with TPMS were asked to answer the questions in the Supplemental Form.

The Tire Pressure Interview Form was used to obtain information regarding the history of the vehicle and its tires, as well as how air is added to the vehicle's tires (e.g., by whom, when, for what reasons). In addition, driver profile data was recorded, as well as driver knowledge about how to keep proper tire pressure and where/how the driver obtained information about tire care and tire pressure.

The Refueling Interview Form was used to obtain information regarding the driver's refueling habits, as well as characteristics of vehicle's occupants (e.g., number, reason for traveling).

The Supplemental Form was used to obtain information regarding the drivers' knowledge of their TPMS (e.g., location and purpose of the warning lamp and the malfunction lamp, how to reset the lamps, and what maintenance, if any, service has been required of the TPMS). Drivers could select one of four ways to complete the questions in the Supplemental Form (i.e., an on-site interview, a mail-back questionnaire, an online questionnaire, and a call-back interview) but almost all drivers who agreed to participate selected the on-site interview.

2.8 Questionnaire Development and Training

Development and Pilot Testing of Data Collection Forms

Data collection forms were developed in consultation with NHTSA subject matter specialists for tires, tire pressure, and fuel economy. Initial testing of the survey forms was completed informally within the agency, with later testing completed by the NASS Zone Centers during a pre-pilot test in May of 2010.

Recruitment of Field Staff

Field data collection was conducted through the infrastructure of the National Automotive Sampling System Crashworthiness Data System (NASS-CDS), which has teams of researchers located at Primary Sampling Units (PSUs) throughout the United States. Members of the TPMS-SS data collection teams were selected from these researchers, many of whom had experience conducting the prior tire pressure studies – the 2001 Tire Pressure Special Study (TPSS) and the 2009 Tire Pressure Monitoring System Study (TPMSS). Each PSU had two or more staff persons who participated in the study in teams of two researchers. One of the researchers, normally the most experienced staff member at the PSU, served as the team leader, and was able to use prior experience to assist other staff members in their data collection. After the data collection on each participant was completed, the team leader was responsible for the survey forms being complete, accurate and legible. In addition, the team leaders were responsible for reviewing collected data against the digital images taken (e.g., the vehicle VIN).

Training and Pilot Testing of the Data Collection Protocols

A PowerPoint preview of the TPMS-SS study was presented to the NASS Zone Centers and PSU staff at the NASS annual training in December of 2009. In 2010, they participated in two different webinar trainings: 1) June training on the application that had been developed for use in entering the TPMS-SS survey data, and 2) July training on the survey forms and data collection protocols. After training was completed, a pilot test of the TPMS-SS survey forms and data collection protocols was conducted in each PSU. Data obtained from the pilot tests were entered into the application, in order to double check the data entry application and the survey forms, as well as provide experience with entering the survey data into the data entry application. In addition, a data collection procedures manual was distributed for reference. Since data collection continued into 2011, another PowerPoint presentation on TPMS-SS was made at the NASS annual training in December of 2010.

Unannounced Site Visits

NHTSA and Zone Center staff paid unannounced site visits to monitor the quality of data collection in the PSUs.

2.9: Data Entry and Quality Control

Data Entry

Data from the seven paper forms used in the survey were entered manually by the data collectors into an application developed specially for the TPMS-SS survey. This data application contained automated edit checks, skip patterns and other features to help insure that the data were entered correctly. In addition, staff at the Zone Centers checked the data that were entered, including checking the images that had been taken.

Quality Control

After the data were entered, checks were run by NHTSA staff to identify outliers, discrepancies between two similar variables, and other such inconsistencies via automated logic checks and data runs. While information about data elements that flagged these edit checks was sent to the Zone Centers to be reviewed and, if necessary, corrected, no statistical editing was performed to alter the recorded values of outliers.

After data reconciliation, a final file was translated into SAS data sets. In addition, database reconciliation of these final SAS data sets was conducted.

2.10: Weighting and Estimation

The PSU-level weights have been previously established by the NASS-CDS and reflect diverse factors such as urbanization and population demographics. The weights at the remaining levels of sampling (ZIP code, gas station, and vehicle) were computed as simple random samples with the final weights representing the inverse of the probability of unit selection in the study.[7] For example, if the PSU had 70 ZIP codes and seven were selected, each selected ZIP code would have been weighted tenfold. However, it should be noted that logistical concerns precluded a purely simple-random selection procedure at certain stages, precluding perfect national representation. The selection of gas stations within PSUs was not fully randomized in order to ensure that selected stations had necessary facilities and adequate vehicle traffic. Similarly, vehicles could only be observed during certain hours of the day and only at gas stations. Practical restrictions such as these were presumed to introduce no bias, but we cannot be certain that this is the case.

Estimation was conducted using SAS survey procedures to specify the experimental design. The resulting estimates can be considered nationally representative during the period of data collection (August 2010 – April 2011). It is important to bear in mind that these estimates represent only the population of vehicles that were eligible for inclusion in the survey (for example, all data collection was conducted between 7 a.m. and 6 p.m.; vehicles that always refuel outside of this interval were not in the sampled population). It is assumed that the general population will not differ substantially from the sampled population.

2.11: Data Corrections and Adjustments

Prior to analysis several corrections were applied to the raw data. Data were collected on gas stations and vehicles that refused to participate in the study for the purpose of a non-response bias analysis, which revealed no strong bias introduced by non-response at the gas station or vehicle levels. No non-response bias adjustments were made to the final weights. A correction to the observed tire pressures was made to account for tire temperature (measured in degrees Fahrenheit at the outside surface of the tire) using the Ideal Gas Law and estimates of average tire volume, as well as average ambient temperatures. This correction took the form of:

Equation 1: Temperature Adjustment to Observed Tire Pressure

$$adjusted\ pressure = observed\ pressure - [(observed\ temp - 65) * .1]$$

This correction was necessary because of the large influence of temperature on tire pressure. This correction standardizes the observed pressures relative to a fixed average ambient temperature (65 degrees Fahrenheit) in order to minimize any potential bias caused by tire temperature. For example, consider a rural gas station near a freeway. It is possible that the vehicles visiting the station will fall into two general categories: those passing through on the freeway that are more likely to have very hot tires and TPMS, and local vehicles that will have colder tires and are less likely to have TPMS. The correction described above prevents any bias that could be introduced by such a situation. Because the

---

[7] There is fairly large variability in the sampling weights generated by the complex sampling design. The smallest observation weight is 22.3, while the largest is 17837.6. The mean of the weights is 998.5 and the standard error of the mean is 13.5. Although the variability is large, a sensitivity analysis using trimmed weights did not yield substantially different results from the untrimmed weights. Based on these results, the untrimmed weights are used for all statistical analyses in the report.

same correction is applied to both the treatment (TPMS) and control (no TPMS) vehicles the exact value of the average ambient temperature chosen should be inconsequential to the effectiveness estimates.

The TPMS-SS includes data from vehicles of model year 2004-2011. Vehicles that did not fall within this range were not approached for inclusion in the survey. In the primary analyses of TPMS effectiveness the vehicle model year range is restricted to 2004-2007 because all vehicles of model year 2008 and newer are equipped with TPMS and from that point on direct comparisons of pre- and post-standard vehicles of the same model year are impossible.

In addition to the preceding adjustments the data were checked for internal consistency, for example by checking to see that the observed type of TPMS agreed with the available options on the specific make and model of vehicle. In instances where variables showed disagreement, the discrepancy was rectified using other collected variables when possible and if this was not possible the conflict was resolved by forcing agreement with the ostensibly more reliable of the two conflicting variables.

3: Underinflation

3.1: Summary

Based on the results of this survey, it is estimated that 71.1 percent of all passenger vehicles of model year 2004-2011 in the United States have at least one tire that is underinflated by at least one psi relative to the recommended pressure, and that 12.4 percent have at least one tire that is severely underinflated (more than 25% below the vehicle manufacturer's recommended cold tire pressure). The percentage of model year 2004-2007 passenger vehicles with at least one tire severely underinflated was 23.1 percent for vehicles without TPMS and 11.8 percent for vehicles with direct TPMS. Based on these rates, TPMS is estimated to be 55.6-percent effective at preventing severe underinflation.

Significant differences were found for different types of TPMS warning displays (single warning lamp, tire-specific warning lamp, or tire-specific real-time pressure readout). However, these differences may be due to a vehicle model bias rather than to genuine differences in effectiveness among the different display types, given that vehicle models with tire-specific and pressure readout displays are likely to be different (e.g., more expensive) than models with a single warning lamp.

TPMS was also found to be about 14 percentage points more effective in preventing severe underinflation in light trucks and vans (LTVs) than in passenger cars (PCs). There were no differences in effectiveness for less severe levels of underinflation by vehicle type.

3.2: Population Estimates

Underinflation is measured at the vehicle level and is defined by the lowest inflation pressure of any tire on a vehicle compared to the manufacturer's recommended cold tire pressure. Under FMVSS No. 138 a TPMS must alert a driver when a tire's pressure is more than 25 percent below the manufacturer's recommended pressure. Underinflation at or below this threshold established in the standard will be referred to in this report as 'severe underinflation'. Because negative safety effects of underinflation could also result from lesser degrees of underinflation, several alternative levels of underinflation will also be considered. Underinflation of 10 percent or more is referred to as 'moderate underinflation', while underinflation greater than 0 percent is referred to as 'any underinflation'. Measurements of tire pressure were taken with precision of one psi.

Analyses of underinflation are done at the vehicle level. The least inflated tire after temperature correction is selected to represent the vehicle. Figure 2 below shows the weighted distribution of this least inflated tire for the 6,103 vehicles in the study from which researchers were able to collect the following information: pressure and temperature of all four tires, make, model, and year of vehicle, and presence and type of TPMS. Once weighted, the vehicles included in Figure 2 below represent 6,108,266 vehicles in the United States. The application of weights accounts for factors such as demographic make-up and geographic region, and result in a nationally representative set of data. The positive range on the right side of the graph represents vehicles on which all four tires are inflated above the recommended level.

Figure 2: Distribution of Least Inflated Tire – All Surveyed Vehicles

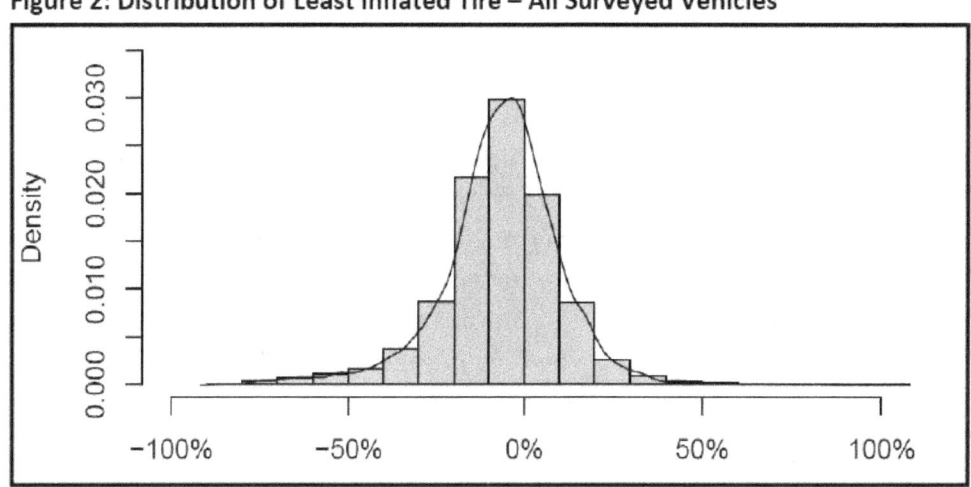

The mean of the distribution shown above is -6.626 percent (6.626 percent below manufacturer's recommended cold pressure) and the median is -5.667 percent. The underinflation rates for several cutoff points along with their standard errors are reported in Table 2 on the following page. The first row in the table reflects the overall distribution of surveyed vehicles shown above. The other rows represent the findings for specific subsets of the survey sample.

Table 2: Weighted Population Underinflation Rates

	Any Underinflation	At Least 10% Below	At Least 25% Below
All Surveyed Vehicles	71.1% (±2.1%)	41.3% (±2.4%)	12.4% (±1.5%)
Vehicle Type			
Passenger Cars	70.0% (±2.2%)	39.9% (±2.4%)	12.2% (±1.5%)
Light Trucks/Vans	72.3% (±2.4%)	42.7% (±2.5%)	12.6% (±1.6%)
TPMS			
No TPMS (MY 2004-2007)	73.5% (±2.4%)	53.9% (±2.7%)	23.1% (±2.2%)
Direct TPMS (MY 2004-2007)	73.7% (±2.6%)	43.1% (±3.1)	11.8% (±2.3%)
Direct TPMS (MY 2008-2011)	68.4% (±3%)	32.2% (±2.7%)	5.7% (±1%)
Direct TPMS (MY 2004-2011)	70.2% (±2.7%)	36% (±2.8)	7.8% (±1.4%)

There is a clear difference in the rates of severe underinflation for surveyed vehicles with and without TPMS. The difference for moderate underinflation is much more modest, and there appears to be little difference for any underinflation. There does not appear to be much of an effect for vehicle type, although LTVs appear to have slightly higher rates of underinflation.

Comparison of these rates with past research is difficult. The Tire Pressure Special Study (2001) was conducted before FMVSS No. 138 was proposed and the population results are reported in absolute psi rather than percent difference between observed and recommended pressure. The Tire Pressure Monitoring System Study (2005) only reported the proportion of vehicles with severely underinflated tires that were TPMS-equipped, not the proportion of vehicles that had severely underinflated tires.

3.3: TPMS Effectiveness

To evaluate the effectiveness of TPMS, the collected survey data were restricted to the model years 2004-2007. These were the only model years that contained vehicles both with and without TPMS and by restricting the analysis to these model years it is less likely that bias was introduced by having substantially different types of vehicles in the comparison groups (vehicles with and without TPMS). It is also likely that there is a correlation between proper inflation rate and model year, particularly for newer vehicles which will be more likely to have properly inflated tires. Restricting included model years should also attenuate any bias introduced by the interaction between the effects of vehicle age on the likelihood of underinflation and the likelihood of the presence of TPMS.

All vehicles with indirect TPMS systems were removed from the overall effectiveness analysis, although they are considered separately in section 3.4. These systems were removed because they were all earlier indirect systems that are not capable of meeting the current standard and are unlikely to represent the effectiveness of either direct systems or the more advanced indirect systems that do meet the requirements of FMVSS No. 138.

The graph below shows the ratio of vehicles surveyed with one or more tires that falls below three different threshold levels of underinflation: more than 0 percent (any underinflation), more than 10 percent (moderate underinflation), and more than 25 percent (severe underinflation). Vehicles are plotted across model year and grouped by presence of TPMS.

Figure 3: Population Underinflation Rates by Model Year and TPMS

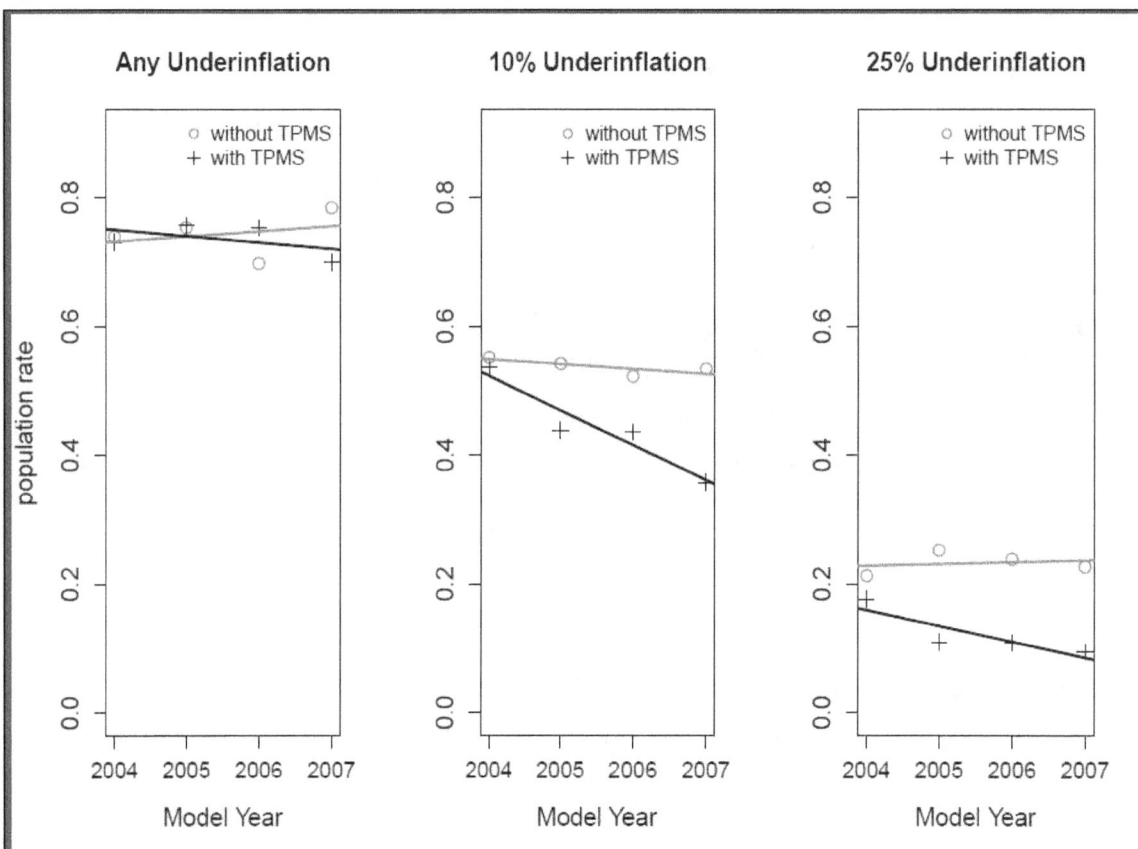

The graphical analysis of the data shows that vehicles with TPMS are less likely to have one or more tires that are moderately or severely underinflated. Notice there is little or no difference between TPMS-equipped and non-equipped vehicles in the left graph showing the rate of "any" underinflation, not a surprising result given that most TPMS systems will only alert a driver to underinflation if it is 25 percent or worse.

Another important finding of these data is that while underinflation rates among vehicles without TPMS seem to be very stable across model year, vehicles with TPMS show a consistent decline in underinflation rates as model year increases. There are several plausible explanations for this observed increase in effectiveness. It is possible that TPMS technology improved from 2004 to 2010 (for example,

Figure 1 shows that direct systems became much more common relative to indirect systems over this period). It is also possible that TPMS suffers from attrition when malfunctioning units are not repaired or when the systems are not recalibrated after sensors are replaced. A third hypothesis is that drivers may pay less attention to the messages from TPMS as the vehicles age. The population rates of underinflation for MY 2004-2010 vehicles equipped with TPMS are shown in Figure 4 below.

Figure 4: Population Underinflation Rates: All Surveyed Vehicles With TPMS

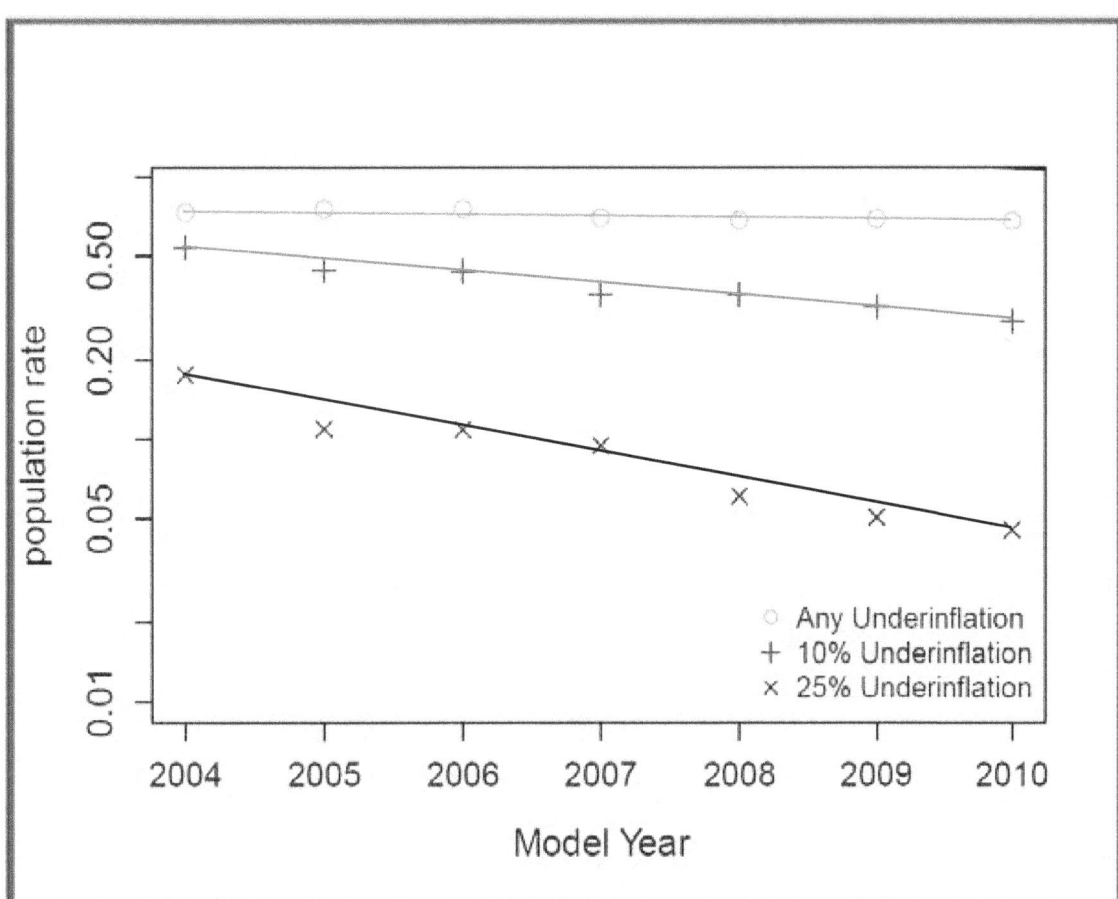

The trend toward decreasing underinflation rates as model year increases seen in Figure 3 clearly continues through 2010, but with the data available it is impossible to determine the underlying causes of this linear trend. By averaging across model years 2004-2007 to create an effectiveness estimate, the resulting estimate will be an accurate depiction of the observed effectiveness in the sampled population of MY 2004-2007 vehicles, but because of the trend it will likely be a conservative estimate (underestimate) of the true effectiveness of adding a TPMS to a new vehicle. Conversely, if the hypotheses of system attrition or reduced driver response are correct, it could be an overestimate of the future effectiveness in older vehicles.

Effectiveness estimates are derived by computing the relative risk of underinflation for vehicles with TPMS versus vehicles without TPMS. This method compares the likelihood that a vehicle without TPMS will have at least one underinflated tire to the likelihood that a vehicle with TPMS will have at least one underinflated tire. If the probability of underinflation is the same for the two groups, then the relative risk will be one and the percent effectiveness will be zero. If the incidence is less for vehicles with TPMS than for vehicles without TPMS, then the relative risk will be less than one. The relevant formulae for risk ratio and percent effectiveness are given below; all statistics used are weighted and corrected for sampling design.

Equation 2: The Risk Ratio

$$risk\ ratio = \left(\frac{\#\ of\ underinflated\ vehs\ w/\ TPMS}{\#\ of\ properly\ inflated\ vehs\ w/\ TPMS}\right) \Big/ \left(\frac{\#\ of\ underinflated\ vehs\ w/o\ TPMS}{\#\ of\ properly\ inflated\ vehs\ w/o\ TPMS}\right)$$

Equation 3: Percent Effectiveness

$$percent\ effectiveness = (1 - risk\ ratio) * 100$$

In the preceding equations 'underinflated' and 'properly inflated' are determined by the threshold of interest. For example, when evaluating effectiveness at preventing underinflation of 25 percent or more, a vehicle whose least inflated tire is 20 percent below the manufacturer's recommended pressure would be considered properly inflated.

Table 3 below gives the effectiveness estimates and their 95-percent confidence intervals for the three underinflation thresholds shown in Figure 3 above. Estimates and confidence intervals are computed using SAS PROC SURVEYFREQ, which accounts for sample design when estimating variance. The percent effectiveness estimates may be interpreted as the reduction in likelihood of underinflation that results when a vehicle is equipped with TPMS.

Table 3: TPMS Effectiveness Estimates (MY 2004-2007)

	Any Underinflation	> 10% Underinflation	> 25% Underinflation
Percent Effectiveness	-1%	35.3%*	55.6%*
95% Confidence Interval	(-45.8%, 29.7%)	(13.4%, 51.6%)	(36%, 69.2%)
*=statistically significant at p<.05			

TPMS is estimated to be 56-percent effective at preventing severe underinflation of 25 percent or more below the manufacturer's recommended cold tire pressure. It is also estimated to be very effective (over 35%) at preventing moderate underinflation of 10 percent or more. It does not appear to have an effect on mild underinflation, which includes any underinflation whatsoever.

3.4: Effectiveness by Display and System Type

The analyses of display type and system type will differ slightly from other analyses in this report because of unequal sample size. Different statistical methods are necessary because of the large inequality in sample sizes that would result from an analysis of overall effectiveness. For example, if vehicles with tire-specific PSI displays were compared to vehicles without TPMS there would be very few of the former and thousands of the latter. This inequality could lead to spurious statistical conclusions, namely that small differences may be found statistically significant because of the large size of the control group (vehicles without TPMS). Instead of using risk ratios the estimates in this section and the following section on PC/LTV effectiveness are rates of underinflation within the analysis groups. This means, for example, that vehicles with tire-specific PSI readouts are compared to standard systems with only a warning lamp, rather than vehicles without TPMS. A logistic regression is then used to determine if these differences are statistically significant. This method of analysis will be less sensitive to unequal sample sizes.

Display Type: Because there are relatively few observations of vehicles with tire-specific warning lamps or tire-specific PSI displays, it is difficult to estimate the overall effectiveness individually for these separate technologies. However, it is possible to gain some insight into relative effectiveness by comparing underinflation rates in all vehicles with TPMS by TPMS display type: warning lamp only, tire-specific warning lamp, and tire-specific PSI display. This allows the entire range of collected model years (2004-2011) to be included because the comparison is made within vehicles with TPMS, not between vehicles with TPMS and vehicles without TPMS.

A one-way ANOVA showed a significant effect of display type on the amount of underinflation observed in vehicles with TPMS (vehicles without any type of TPMS or with indirect TPMS were removed from this model). Table 4 below shows the weighted severe underinflation rates for the different TPMS display technologies.

Table 4: Rate of Severe Underinflation by TPMS Display Type (MY 2004-2011)

TPMS Display	Rate of Severe Underinflation	95% Confidence Interval
Warning Lamp Only	7.93%	(5.04%, 10.81%)
Tire-Specific Warning	11.67%	(4.83%, 18.51%)
Tire-Specific PSI	5.67%	(2.71%, 8.63%)

Vehicles with tire-specific inflation pressure displays have the lowest rate of severe underinflation. Unexpectedly, vehicles with tire-specific warning lamps appear to have a higher rate of severe underinflation than standard systems with a single underinflation warning lamp. A logistic regression designed to test the significance of this difference while controlling for vehicle age and vehicle type (PC or LTV) showed that both differences are statistically significant. Based on the percentages in the table above systems with tire-specific PSI readouts are 40 percent more effective than standard warning lamp only systems at preventing severe underinflation, while systems with tire-specific lamps are 32 percent less effective. These differences are statistically significant ($p = .004$ and $p = .007$, respectively).

It is important to note that in the primary analysis of overall TPMS effectiveness, every effort was made to ensure that the vehicles being compared were similar to one another. This was done by restricting the model year range of vehicles, and the assumption of vehicle model homogeneity was checked by restricting the makes and models included in a supplemental analysis to only vehicles that were produced both with and without TPMS (see section 5.2: Possible Sources of Bias). The same efforts could not be made here because of the limited sample size and as a result, the types of vehicles being compared may be dissimilar and this may be driving the observed differences rather than the type of TPMS display. Vehicle models with tire-specific warning lamps and PSI displays are likely to be more similar to one another (more expensive, more performance-oriented, etc.) than to vehicle models with only a single warning lamp. Until more data can be collected on vehicles with tire-specific TPMS displays it will be difficult to separate differences due to vehicle model from differences due to TPMS display type.

System Type: In this survey the only observed vehicles with indirect TPMS (n=209) were produced prior to FMVSS No. 138, and therefore the following analysis is restricted to model years 2004-2007. More recent indirect systems that meet the requirements of FMVSS No. 138 may have similar effectiveness to direct systems, but this will not be possible to verify until there are enough vehicles on the road to provide supporting data.

Table 5: Rate of Severe Underinflation by System Type (MY 2004-2007)

System Type	Rate of Severe Underinflation	95% Confidence Interval
Direct	11.78%	(7.06%, 15.52%)
Pre-FMVSS No. 138 Indirect	15%	(7.57%, 22.42%)

The direct systems in the survey have a slightly lower weighted rate of severe underinflation than the indirect systems. Based on these observed rates, severe underinflation is 21 percent less common in vehicles with direct systems than in vehicles with indirect systems. Both a Chi-square analysis of the unweighted data and a logistic regression of the weighted data gave similar results (p=.09) that suggest that this difference is marginally significant. It should be noted that these results do not apply to more recent advanced indirect systems that meet the requirements of FMVSS No. 138.

3.5: Effectiveness by Vehicle Type

As noted earlier in this report, there is a linear effect on the effectiveness of TPMS over time, with newer TPMS systems showing lower underinflation rates than older TPMS systems (see Figure 4). If there is a linear effect of time on the ratio of passenger cars (PCs) to light trucks and vans (LTVs) as well as a large difference in underinflation rates between these two groups, then this could create a source of bias in the overall effectiveness estimates. Although Table 2 shows that in the overall sample there is a negligible difference in underinflation rates for PCs and LTVs, a larger difference may exist in the older

2004-2007 vehicles included in the effectiveness analysis which could result in a directional bias. This analysis may also shed some light on the linear trend in effectiveness of TPMS systems over time illustrated by Figure 4. Figure 5 below shows the relative frequency of PCs and LTVs by model year in the vehicles observed by the TPMS-SS survey.

Figure 5: PC/LTV Frequencies by Model Year

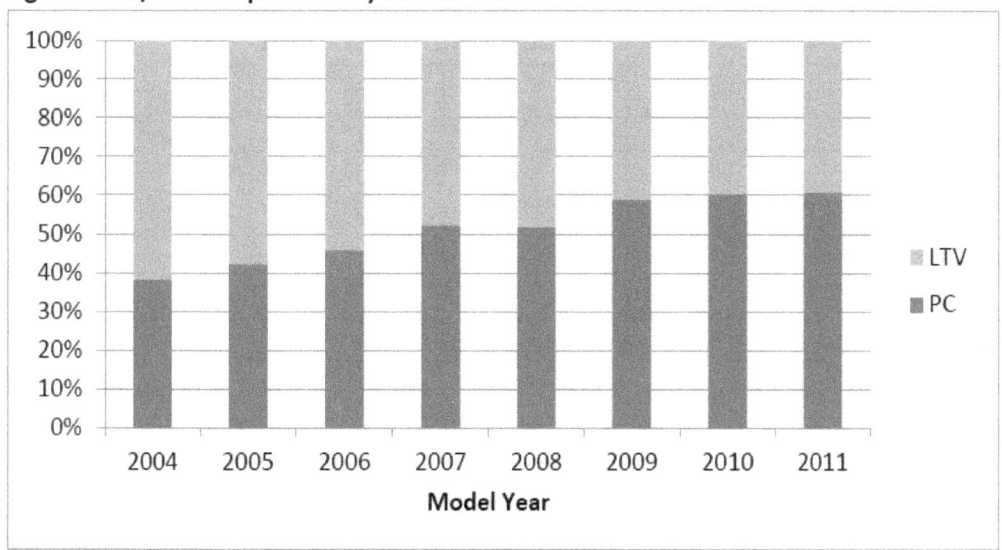

There appears to be a moderate trend towards a higher proportion of passenger cars relative to LTVs as model year increases. The trend appears especially strong during the model years included in the effectiveness evaluation, 2004-2007. This trend towards relatively fewer LTVs is also found in vehicle registration data from R.L. Polk. According to Polk, registrations in calendar year 2010 ranged from 46 percent PCs in model year 2004 to 51 percent PCs in model year 2007. These numbers are very similar to those observed by TPMS-SS collection staff.

Effectiveness by vehicle type was evaluated using the same statistical methods as the overall evaluation of effectiveness (see section 3.3: TPMS Effectiveness). Table 6 below shows the effectiveness of TPMS at preventing severe underinflation by vehicle type. The model years included (2004-2007) are the same as for the overall analysis.

Table 6: TPMS Effectiveness by Vehicle Type (MY 2004-2007)

Passenger Cars	Any Underinflation	> 10% Underinflation	> 25% Underinflation
Percent Effectiveness Of TPMS	14.2%	33.7%*	47.3%*
95% Confidence Interval	(-27%, 42%)	(17.6%, 46.7%)	(22.7%, 64.1%)
Light Trucks/Vans			
Percent Effectiveness Of TPMS (95% CI)	-14.5%	37.7%*	61.2%*
95% Confidence Interval	(-111.8%, 38.1%)	(3.3%, 59.9%)	(39.9%, 75%)
*=statistically significant at p<.05			

There is a modest difference in effectiveness for severe underinflation between vehicle types, with LTVs showing higher effectiveness (61.2%) than passenger cars (47.3%). Because the relative rate of LTVs is decreasing over time, this difference can't be responsible for the observed increase in TPMS effectiveness as model year increases. This moderate difference coupled with the moderate change in fleet composition over the included model year period should not be introducing any appreciable bias to the overall estimates of TPMS effectiveness, although it suggests that the linear effect of time on TPMS effectiveness may be even stronger than suggested by Figure 4, which combines PCs and LTVs.

4: Overinflation

4.1: Summary

Although overinflation poses less of an established safety risk than underinflation, it may still have negative effects on vehicle stability and tire integrity, wear and traction. These negative effects may be accompanied by a non-safety benefit of improved fuel economy. The results of this analysis suggest that TPMS is 30.7-percent effective at preventing severe overinflation (more than 25% above the manufacturer's recommended cold tire pressure). There were no significant differences in effectiveness for different TPMS display types, but LTVs showed a significantly higher effectiveness rate than PCs. The use of 25 percent as a threshold value to define severe overinflation is a bit more arbitrary here than in the case of underinflation, where research has been conducted on the effects of various levels underinflation. In the absence of research capable of informing a more meaningful threshold it is used throughout this section as the point that defines severe overinflation.

4.2: Population Estimates

Like underinflation, overinflation is defined at the vehicle level and is analyzed by selecting the most inflated tire on a vehicle to represent that vehicle. The same inclusion criteria that were used in the analysis of underinflation were applied to the data, resulting in inclusion of the same 6,103 vehicle observations. Figure 6 below shows the distribution of the most inflated tire of each of the vehicles included. Percent values are relative to the manufacturer's recommended cold tire pressure and have been corrected to reflect the temperature of the tire at the time of measurement. Note that it is possible for a single vehicle to be simultaneously severely underinflated and severely overinflated.

Figure 6: Distribution of Most Inflated Tire – All Surveyed Vehicles

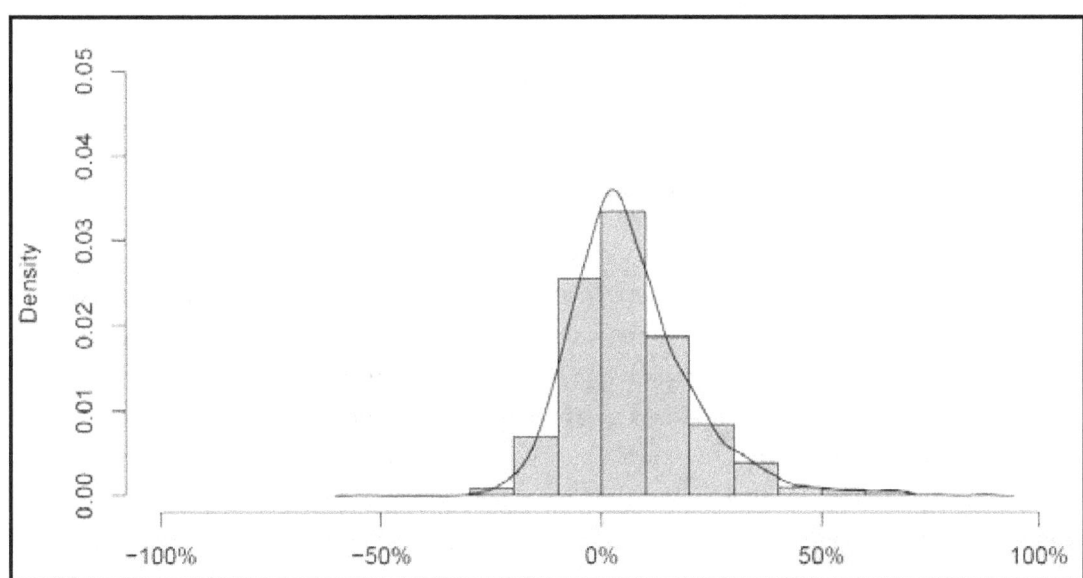

Overinflation seems less prevalent than underinflation in general, although most vehicles have at least one tire that is above the manufacturer's recommended cold tire pressure. The mean of the distribution is 6.7 percent and the median is 4.7 percent. The overinflation rates for several cutoff points along with their standard errors are reported in Table 7 below.

Table 7: Population Overinflation Rates (All Surveyed Vehicles)

	Any Overinflation	At Least 10% Above	At Least 25% Above
All Surveyed Vehicles	**58.1%** (±2.5%)	**34.3%** (±2.2%)	**11.9%** (±1.3%)
Vehicle Type			
Passenger Cars	**60.9%** (±2.7%)	**35.6%** (±2.7%)	**12.7%** (±1.7%)
Light Trucks/Vans	**55.9%** (±2.6%)	**33.2%** (±2.0%)	**11.3%** (±1.3%)
TPMS			
No TPMS	**58.0%** (±2.7%)	**35.6%** (±2.2%)	**13.5%** (±1.7%)
TPMS	**58.3%** (±2.7%)	**32.6%** (±2.6%)	**10.0%** (±1.3%)

These results show that severe overinflation is only slightly less common than severe underinflation. There are probably several different factors that are contributing to the high incidence of overinflation. There may be unintentional overinflation due to confusion between the recommended tire pressure posted on the placard inside the driver's door jamb and the maximum pressure posted on the sidewall of each tire or because people are filling their tires by eye without an accurate gauge. There may also be some intentional overinflation in order to improve fuel economy. Drivers might also intentionally put a little extra air in their tires to assure themselves the tires are not underinflated and to reduce how often they need to check or refill the tires. Based on the high observed rate of overinflation it may be worthwhile to investigate more closely the effects of overinflation on safety and tire integrity, function and wear.

4.3: TPMS Effectiveness

For the evaluation of TPMS effectiveness in preventing overinflation the included model years were restricted to 2004-2007. These model years include vehicles both with and without TPMS, allowing for a direct comparison.

Figure 7 below shows the ratio of vehicles surveyed with one or more tires that fall above three different threshold levels of overinflation: more than 0 percent, more than 10 percent, and more than 25 percent. Vehicles are grouped by model year and by presence of TPMS.

Figure 7: Overinflation Rates by Model Year and TPMS

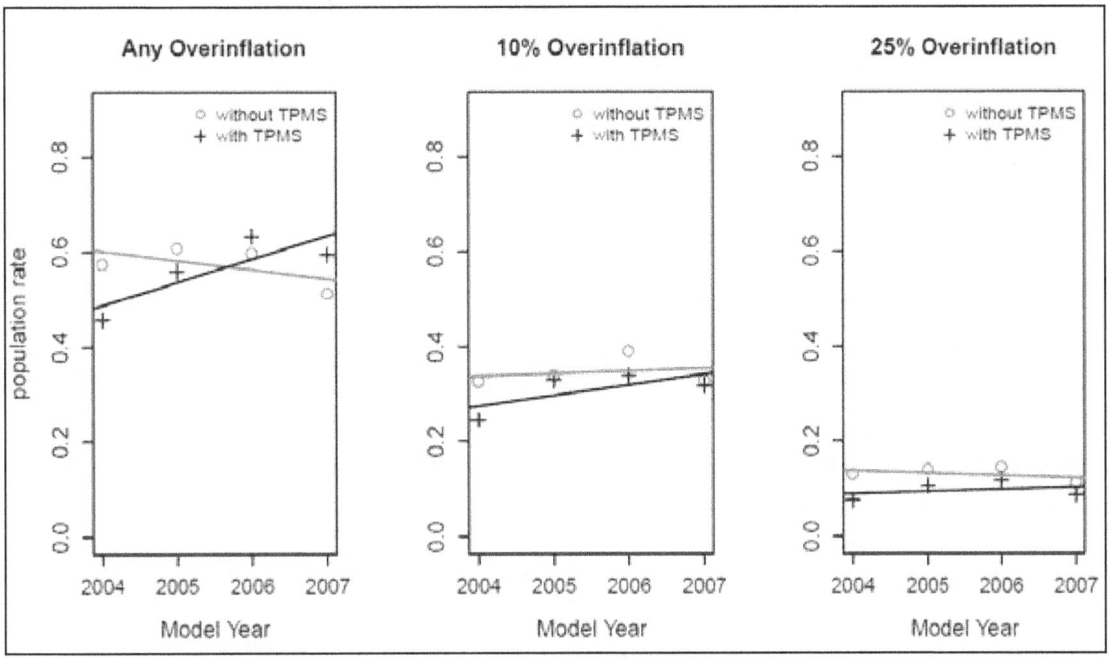

TPMS appears to have no consistent effect on the rate of any overinflation, but some effects on the rates of moderate and severe overinflation. In order to quantify these effects and determine if they are statistically significant, the same method of statistical analysis used for underinflation (risk ratios and associated percent effectiveness) was employed here.

Table 8: TPMS Effectiveness Estimates (MY 2004-2007)

	Any Overinflation	> 10% Overinflation	> 25% Overinflation
Percent Effectiveness	-1.5%	13.3%	30.7%*
95% Confidence Interval	(-22%, 15.6%)	(-1.7%, 26.1%)	(4.9%, 49.5%)
*=statistically significant at p<.05			

This analysis shows that TPMS is significantly effective at preventing severe overinflation. Although there is little research into the specific consequences of overinflation, it is likely that overinflation will have negative effects on vehicle handling and stability, as well as tire wear and integrity. The decline in overinflation attributed to TPMS should be considered a positive associated and unintended benefit of the technology.

It is difficult to explain how TPMS systems prevent overinflation. Some TPMS (11.5% of the surveyed vehicles with TPMS) give a real-time pressure reading of all four tires as a dash display, and these systems would inform the driver of overinflation, although without an associated warning light. But the effectiveness was found to be equally large if not larger for the far more common systems (88.5% of the surveyed vehicles with TPMS) that do not report pressure information and only alert the driver to underinflation through a warning light. It is possible that TPMS encourages good tire maintenance in general which results in reduced overinflation as well as underinflation. Drivers who habitually overinflate their tires as a hedge against underinflation and frequent checking might perhaps be less inclined to do so when they know that TPMS will warn them of an underinflation problem.

4.4: Effectiveness by Display and System Type

Display Type: The same methods used in the corresponding section within the underinflation analysis were employed to estimate the relative effectiveness of the different TPMS display types observed in the survey. An ANOVA did not show any difference in effectiveness on general overinflation for different display types. The same logistic model used in the underinflation analysis was applied here to check for any differences in effectiveness on severe overinflation. This analysis also failed to find any significant effect of display type, vehicle type, or model year on overinflation rate.

Although it was hypothesized that TPMS with a tire-specific PSI display would be driving the observed effectiveness in overinflation reduction, this does not appear to be the case. The rates of severe overinflation by display type are shown in Table 9 below.

Table 9: Rate of Severe Overinflation by TPMS Display Type (MY 2004-2011)

TPMS Display	Rate of Severe Overinflation	95% Confidence Interval
Warning Lamp Only	10.53%	(7.41%, 13.65%)
Tire-Specific Warning	6.7%	(1.64%, 11.77%)
Tire-Specific PSI	11.94%	(5.5%, 18.37%)

Notice that the tire-specific PSI displays actually have the highest observed rate of severe overinflation. Because the differences are not statistically significant one should avoid drawing any conclusions from these differences.

System Type: Exploratory analysis did not suggest that there was a significant effect of system type on overinflation rates. The rates for direct and indirect systems are reported in Table 10 on the following page.

Table 10: Rate of Severe Overinflation by System Type (MY 2004-2007)

System Type	Rate of Severe Overinflation	95% Confidence Interval
Direct	11.94%	(9.27%, 14.61%)
Indirect	14.99%	(8.37%, 21.61%)

The small observed difference in severe overinflation rates is not statistically significant.

4.5: Effectiveness by Vehicle Type

TPMS effectiveness by vehicle type was estimated using the methods described in section 3.3: TPMS Effectiveness. The results are shown in Table 11 below.

Table 11: TPMS Effectiveness by Vehicle Type (MY 2004-2007)

Passenger Cars	Any Overinflation	> 10% Overinflation	> 25% Overinflation
Percent Effectiveness of TPMS	-8.3%	11.4%	12.9%
95% Confidence Interval	(-29.1%, 9.2%)	(-21.9%, 35.5%)	(-44.6%, 47.5%)
Light Trucks/Vans			
Percent Effectiveness of TPMS	1.4%%	14.3%	42.6%*
95% Confidence Interval	(-33%, 26.9%)	(-18.8%, 38.2%)	(16.3%, 60.6%)
*=statistically significant at p<.05			

The only significant result is the reduction in severe overinflation in light trucks and vans. Although these findings help to pinpoint where the benefits are taking place, they do not shed much light on how TPMS is preventing overinflation. It is also interesting to note that effectiveness increases at the higher levels of overinflation. TPMS will not alert a driver to any level of overinflation, and one would expect that if the observed effectiveness is caused by driver characteristics then the effect might be about the same at all levels of overinflation severity.

5: Fuel Economy

5.1: Summary

Although the primary goal of TPMS is to reduce underinflation in order to make vehicles safer to operate, a further benefit of reduced underinflation is improved fuel economy. By combining estimates of reduced underinflation due to TPMS with estimates of increases in fuel economy resulting from increases in tire pressure, it's possible to estimate the amount of fuel that TPMS will save an average vehicle during a given period of time. During the first eight years of operation TPMS is estimated to save a typical passenger car 9.32 gallons of fuel and typical LTV 27.89 gallons of fuel. During 2011 TPMS is estimated to have saved $511,066,488 across the vehicle fleet.

5.2: Vehicle-Level Average Underinflation

While the least and most inflated tires on a vehicle are key metrics in terms of vehicle safety, it is important from a fuel economy standpoint to consider the average underinflation per tire across the four tires on a vehicle. The vehicle-level average underinflation is shown in Table 12 below. Underinflation is shown as a negative number and overinflation as a positive number. For example, for a vehicle with two tires underinflated by three PSI each and the other two overinflated by one PSI each, the average underinflation would be [(-3) + (-3) + (1) + (1)]/4 = -1, or underinflated by an average of 1 PSI per tire. This is a meaningful number because of the approximately linear relationship between tire pressure and fuel economy over the inflation range that was observed in the TPMS-SS.

Average underinflation was computed for every complete vehicle observation in the TPMS-SS and averaged across vehicle model year. The results for vehicles with and without TPMS are shown in Figure 8 below.

Figure 8: Vehicle Average Underinflation

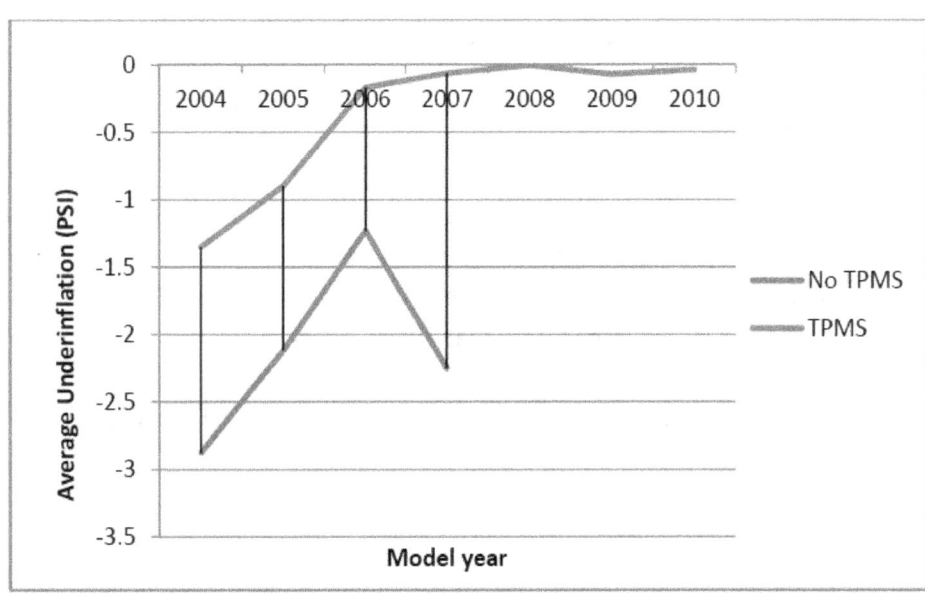

Although TPMS was shown in the previous section to result in decreases in overinflation as well as underinflation, vehicles with TPMS show less average underinflation than vehicles without TPMS. The difference is fairly consistent, except for the large drop in 2007 vehicles without TPMS. This data point represents the smallest sized group on the figure; by model year 2007 most vehicles were equipped with TPMS. The small sample size may be creating some instability in the estimate, and this graph suggests that it would be prudent to average across model years 2004-2007 to obtain an accurate estimate of the effect of TPMS on average underinflation.

When averaged across model years and vehicle types the difference in average underinflation per tire between vehicles with and without TPMS is 1.05 psi. The difference for passenger cars is 0.714 psi and for LTVs the difference is 1.405 psi. The estimated average total underinflation for vehicles with and without TPMS is shown in Table 12 below with the 95-percent confidence interval.

Table 12: Average Underinflation in PSI

Population	Average Underinflation	95% Confidence Interval
All Surveyed Vehicles	-0.484	(-1.031, 0.062)
PC	-0.179	(-0.673, 0.316)
LTV	-0.736	(-1.394, -0.078)
Vehicles Without TPMS (MY 2004-2007)	-1.406	(-2.252, -0.56)
PC	-0.79	(-1.396, -0.184)
LTV	-1.948	(-3.174, -0.723)
Vehicles With Direct TPMS (MY 2004-2007)	-0.35	(-1.088, 0.388)
PC	-0.076	(-0.785, 0.632)
LTV	-0.543	(-1.456, 0.371)

There is, on the average, less underinflation per tire in TPMS-equipped vehicles because TPMS is quite effective in reducing severe underinflation. In computation of the average effect, this reduction in underinflation is partly attenuated because TPMS also reduces severe overinflation. Therefore the net effect is not as large as it would have been if TPMS only reduced underinflation and had no effect on overinflation.

5.3: Effect of TPMS on Fuel Economy

Tire Pressure and Fuel Economy: Several studies have been conducted to compare the relationship between tire pressure and fuel economy. Table 13 below is adapted from the report 'NHTSA Tire Fuel Efficiency Consumer Information Program Development'[8] and shows the results of several recent studies.

Table 13: Past Research on Tire Pressure and Fuel Economy

Study	Predicted Increased % MPG/PSI Increase
Wicks and Sheets[9]	0.38
U.S. DOE[10]	0.3
Clark et al.[11]	0.1-0.6†
Hall and Moreland[12]	0.1-0.4†
Continental Tire[13]	0.16-.032†
U.S. EPA[14]	.33†
Average	**0.308**
†= mpg calculated through rolling resistance	

Although the analysis methods differed, these reports show strong agreement on the effect of tire pressure on fuel economy and the estimates reported are very similar. On average, fuel economy is expected to increase by 0.308 percent for every 1-percent increase in average tire pressure. Combining this average estimate with the results from Table 12 above allows an estimate of the improvement in fuel economy attributable to TPMS.

---

[8] Evans, Larry R., et al. (2009) *NHTSA Tire Fuel Efficiency Consumer Information Program Development: Phase 2- Effects of Tire Rolling Resistance Levels on Traction, Treadwear, and Vehicle Fuel Economy.* (Report No. DOT HS 811 154), Washington, DC, National Highway Traffic Safety Administration

[9] Wicks, F. & Sheets, W. (1991). "Effect of Tire Pressure and Performance Upon Oil Use and Energy Policy Options." Proceedings of the 26th Intersociety Energy Conversion Engineering Conference IECEC-91, Aug. 4-9, 1991, Boston, Mass. Vol. 4, pp 307. La Grange, IL: American Nuclear Society.

[10] EPA. (2009). Fuel Economy Guide. Washington, DC: Environmental Protection Agency. [Website]. http://www.fueleconomy.gov/feg/maintain.shtml.

[11] Clark, S. K. & Dodge, R. N. (1979). A Handbook for the Rolling Resistance of Pneumatic Tires. Prepared for the U.S. DOT. Ann Arbor, MI: Regents of the University of Michigan.

[12] Hall, D.E. & Moreland, J.C. (2000). Fundamentals of Rolling Resistance. Spring 2000 Education Symposium No. 47, Basic Tire Technology: Passenger and Light Truck. Akron, OH: American Chemical Society.

[13] Continental Tire. (2008). Government Regulation in Transition-Continental Tire Point of View. Presentation before the California Energy Commission.

[14] Grugett, B.C., Martin, E. R., & Thompson, G.D. (1981). The Effects of Tire Inflation Pressure on Passenger car Fuel Consumption. International Congress and Exposition, Feb. 23-27, 1981. Paper #810069, SAE Technical Paper Series. Warrendale, PA: Society of Automotive Engineersm Inc.

Equation 4: Effect of TPMS on Fuel Economy

$$Effectiveness\ (\Delta) = (p_1 - p_0) * \bar{\gamma}$$

Where p_1 is the average underinflation of model year 2004-2007 vehicles with direct TPMS, p_0 is the average underinflation of 2004-2007 vehicles without TPMS (both taken from Table 12 above), and $\bar{\gamma}$ is the average predicted percent increase in fuel economy accompanying a one psi increase in average tire pressure (taken from Table 13 above). Solving separately for passenger cars and LTVs gives:

$$\Delta^{LTV} = -0.543 - (-1.948) * 0.308\% = 0.433\%$$

$$\Delta^{PC} = -0.076 - 0.79 * 0.308\% = 0.22\%$$

Although the estimated improvements in fuel economy are modest, over time and across the fleet they may represent a considerable savings.

Estimated Fuel Savings Per Vehicle: Using some additional estimates and assumptions it is possible to calculate the amount of fuel saved by TPMS over the life of a new vehicle. This quantity is estimated by calculating the putative fuel economy of a hypothetical model year 2011 vehicle without a TPMS system and comparing that fuel economy to the known average fuel economy of a 2011 vehicle (all of which are equipped with TPMS).

In Table 14 below the survival probability and average miles travelled are reported for the first eight years of a new vehicle's life. Only the first eight years are included because as a vehicle ages it becomes more likely that its TPMS will be nonfunctioning due to either mechanical failure (for example, all direct pressure sensors are equipped with a battery) or some other factor. Placing an upper limit on the effective lifespan of a TPMS unit should help prevent an overestimation of the benefits provided during a vehicle's lifetime.

All estimates in this chapter are calculated separately for passenger cars and LTVs due to differences between these two groups in TPMS effectiveness and other characteristics.

Table 14: Estimated Survival Probability and Miles Traveled by Vehicle Age

Passenger cars		
Vehicle Age (Years)	Survival Probability $[P(s)^{PC}]$	Miles Traveled (x^{PC})
1	0.995	14,231
2	0.990	13,961
3	0.983	13,669
4	0.973	13,357
5	0.959	13,028
6	0.941	12,683
7	0.919	12,325
8	0.892	11,956
Light Trucks and Vans		
Vehicle Age (Years)	Survival Probability $[P(s)^{LTV}]$	Miles Traveled (x^{LTV})
1	0.974	16,085
2	0.960	15,782
3	0.942	15,442
4	0.919	15,069
5	0.891	14,667
6	0.859	14,239
7	0.823	13,790
8	0.783	13,323

The survival probabilities[15] and average miles traveled[16] are used in the following equation to estimate the average amount of fuel saved per vehicle. Notice that the summation indices in Equation 4 below are from two to eight; an assumption of this analysis is that TPMS will have no effect on tire pressure for the first year of operation. The survey data showed very little underinflation in vehicles that were produced in the same year as data collection.

---

[15] The survival rates were calculated from R.L. Polk, National Vehicle Population Profile (NVPP), 1977-2003; see NHTSA, "Vehicle Survival and Travel Mileage Schedules," Office of Regulatory Analysis and Evaluation, NCSA, January 2006, pp. 9-11, Docket No. 22223-2218. Polk's NVPP is an annual census of passenger cars and light trucks registered for on-road operation in the United States as of Jul 1 each year. NVPP registration data from vehicle model years 1977 to 2003 were used to develop the survival rates reported in Table VIII-6. Survival rates were averaged for the five most recent model years for vehicles up to 20 years old, and regression models were fitted to these data to develop smooth relationships between age and the proportion of cars or light trucks surviving to that age.

[16] NHTSA, "Vehicle Survival and Travel Mileage Schedules," Office of Regulatory Analysis and Evaluation, January 2006, pp. 15-17 (Docket NHTSA-2009-0062-0012.1). The original source of information on annual use of passenger cars and light trucks by age used in this analysis is the 2001 National Household Travel Survey (NHTS), jointly sponsored by the Federal Highway Administration, Bureau of Transportation Statistics, and National Highway Traffic Safety Administration.

Equation 5: Gallons of Fuel Saved per Vehicle

$$estimated\ gallons\ saved\ (\delta) = \sum_{i=2}^{8} \left[\frac{P(s)_i * x_i}{\hat{\theta}} - \frac{P(s)_i * x_i}{\theta} \right]$$

Where $P(s)_i$ is the probability of survival for a vehicle of age i, x_i is the average number miles traveled by a vehicle of age i, and θ is the estimated on-road fuel economy of a new (2011) vehicle with TPMS, calculated by combining the average fuel economy for a passenger car (33.8 MPG) or LTV (24.5 MPG)[17] with the correction factor employed regularly by the DOT to account for the discrepancy between estimated fuel economy as reported by the EPA and the observed on-road fuel economy (this factor accounts for typical driver behavior and is equal to 0.8)[18].

Equation 6: 2011 Fuel Economy With TPMS

$\theta^{PC} = 33.8 * 0.8 = 27.04\ MPG$

$\theta^{LTV} = 24.5 * 0.8 = 19.6\ MPG$

The adjusted fuel economy $\hat{\theta}$ is the estimated economy for a hypothetical new vehicle without TPMS, and is calculated with the following equation:

Equation 7: Estimated 2011 Fuel Economy Without TPMS

$\hat{\theta}^{PC} = \theta^{PC} * [1 - (\Delta^{PC})] = 27.04 * (1 - 0.0022) = 26.98\ MPG$

$\hat{\theta}^{LTV} = \theta^{LTV} * [1 - (\Delta^{LTV})] = 19.6 * (1 - .00433) = 19.52\ MPG$

The equations above adjust the fuel economy of a PC or LTV with TPMS by removing the benefit estimated to result from TPMS. Substituting and solving for Equation 5 gives:

$\delta^{PC} = 9.32$ gallons/vehicle

$\delta^{LTV} = 27.89$ gallons/vehicle

Light trucks and vans are estimated to see a much larger benefit in fuel savings than passenger cars over their lifetimes. This is a result of higher TPMS effectiveness, poorer average fuel economy, and more miles travelled on average for LTVs. It is important to note that the methods described preclude any estimation of variance for these quantities because they incorporate several estimates derived by secondary sources that do not report respective variance estimates. It is also important to note that the effectiveness estimates for TPMS that are applied to model year 2011 vehicles are derived from 2004-2007 model year vehicles. The results shown in Figure 8 support this assumption, although there is no

---

[17] Research and Innovative Technology Administration (2012). *Average Fuel Economy of US Light Duty Vehicles*. Washington, DC; Bureau of Transportation Statistics. [Website]. www.bts.gov/publications/national_transportation_statistics/html/table_04_23.html

[18] Parsons, G. G. (1986). *Fuel Economy and Annual Travel for Passenger Cars and Light Trucks: National On-Road Survey*. (Report No. DOT HS 806 971), Washington, DC, National Highway Traffic Safety Administration

way to directly test its validity. The dollar value of these estimated lifetime fuel savings depends on fuel prices, but it is likely to account for a substantial portion of the consumer's cost of a new TPMS, which was estimated in 2007 dollars to range from $100.00 to $202.30 for a passenger car and from $135.62 to $234.55 for an LTV.[19]

Estimating Total Fleet Fuel Savings During 2011: A second estimate of interest is the total amount of money saved during the year of survey data collection (2011) due to reduced fuel consumption across the fleet. In order to estimate this quantity, some additional information is needed. The first is the percentage of vehicles eight years of age or less that were equipped with TPMS by model year. The survey data itself gives this estimate for the model years of interest, 2004-2011. A second necessary estimate is the number of vehicles on the road in 2011 of model years 2003-2011. These numbers can be derived from sales data provided by Ward's and the survival probabilities listed in Table 14 above. Finally, the average fuel economy by model year is required.[20] This number needs to be corrected because it is an average that includes vehicles with and without TPMS for the model years 2004-2007. The true quantity of interest is the average fuel economy of a vehicle with TPMS, which can be calculated using the rate of TPMS in a given model year and the estimate of TPMS effectiveness. In the previous calculations no correction was necessary because only the fuel economy of 2011 vehicles was considered, and all vehicles in 2011 were equipped with TPMS. The correction is made as described in Equation 7 below.

Equation 7: Average MPG Correction

$$\overline{MPG} = (r * \theta) + [(1 - r) * \hat{\theta}]$$

Where $\overline{MPG}$ is the average EPA corrected fuel economy across vehicles with and without TPMS, r is the percent of vehicles equipped with TPMS, θ is the average fuel economy of vehicles with TPMS, and $\hat{\theta}$ is the average fuel economy of a vehicle without TPMS. We wish to solve for θ; however, with two unknowns ($\hat{\theta}$ is also unknown) an additional piece of information is necessary. We can use the estimated TPMS effectiveness derived earlier to describe the relationship between MPG_0 and MPG_1 and thereby arrive at a solvable system of equations.

Equation 8: Fuel Economy With and Without TPMS

$$\theta = \hat{\theta} * (1 + \Delta)$$

---

[19] These estimates do not reflect any maintenance or repair costs, only the cost of a new system. Ludtke and Associates (undated) *Cost, Weight Analysis of Tire Pressure Monitoring Systems*. (Docket No. NHTSA-2011-0066-0003). Washington, MI: Ludtke and Associates.

Table 15 below shows these additional required estimates for PCs and LTVs separately.

Table 15: TPMS Share, Sales, and Fuel Economy by Model Year

Passenger Cars					
Vehicle Age (Years)	Model Year	% With TPMS (r^{PC})	New Vehicle Sales (N^{PC})	Average MPG ($\overline{MPG}^{PC}$)	Average MPG With TPMS (θ^{PC})
1	2011	100	Unk	27.04	27.04
2	2010	100	5,635,433	27.12	27.12
3	2009	100	5,400,890	26.32	26.32
4	2008	100	6,769,107	25.20	25.20
5	2007	66.15	7,562,334	24.96	24.98
6	2006	38.29	7,761,592	24.08	24.11
7	2005	23.79	7,659,983	24.24	24.28
8	2004	29.48	7,482,555	23.60	23.64
Light Trucks and Vans (LTVs)					
Vehicle Age (Years)	Model Year	% With TPMS (r^{LTV})	New Vehicle Sales (N^{LTV})	Average MPG ($\overline{MPG}^{LTV}$)	Average MPG With TPMS (θ^{LTV})
1	2011	100	Unk	19.60	19.60
2	2010	100	6,136,787	20.16	20.16
3	2009	100	5,200,478	19.84	19.84
4	2008	100	6,724,058	18.88	18.88
5	2007	77.19	8,897,981	18.48	18.50
6	2006	44.32	9,287,389	18.00	18.04
7	2005	42.61	9,784,346	17.68	17.72
8	2004	27.27	9,816,018	17.20	17.25

Although the new vehicle sales volume for 2011 was unknown at the time of publication, because there is assumed zero benefit from TPMS for the first year of operation, this is not crucial information for estimation. Equation 7 below calculates the amount of money saved by TPMS in 2011 by estimating the number of gallons of fuel saved that year by vehicles of model years 2004-2010 and then multiplying that estimate by the average price of a gallon of fuel in 2011 ($3.576).[20]

Equation 9: Dollars Saved in 2011

$$estimated\ dollars\ saved\ in\ 2011 = \left[\frac{N_i * r_i * P(s)_i * x_i}{\hat{\theta}_i} - \frac{N_i * r_i * P(s)_i * x_i}{\theta_i}\right] * \$3.576$$

Where N_i is the number of vehicles sold of model year i, r_i is the survival probability to 2011 of a vehicle of model year i, θ_i is the average EPA adjusted fuel economy of a vehicle of model year i, and $\hat{\theta}_i$ is the

---

[20] US Energy Information Administration, *Annual Retail Gasoline and Diesel Prices*. Website: http://www.eia.gov/dnav/pet/pet_pri_gnd_dcus_nus_a.htm.

estimated fuel economy of a vehicle of model year *i* without TPMS. Solving this equation for passenger cars and LTVs separately then summing gives:

$$dollars\ saved\ in\ 2011 = 142{,}915{,}684.6\ gallons * \$3.576 = \$511{,}066{,}488.$$

TPMS is estimated to have saved over 511 million dollars in 2011, and *ceteris paribus* this estimate should increase in future years as more vehicles without TPMS are retired and replaced by vehicles with TPMS. Had every vehicle in model years 2004-2011 been equipped with TPMS, the total dollars saved in 2011 would have been:

$$(dollars\ saved\ in\ 2011 | 100\%\ TPMS) = 218{,}378{,}259.9\ gallons * \$3.576 = \$780{,}920{,}657.$$

This estimate should more accurately reflect the future expected savings resulting from TPMS after the technology has saturated the fleet, assuming an effective TPMS lifespan of eight years and not considering any possible additional maintenance expense during this time.

6: Discussion

6.1: Summary of Results

This analysis estimates that TPMS is 55.6-percent effective at preventing severe underinflation as defined in FMVSS No. 138. As a result, vehicles with TPMS should be less likely to be involved in crashes related to tire failure, long stopping distances, or loss of traction. Furthermore, these vehicles should get better gas mileage and should have tires that last longer and provide better response and handling. TPMS is also estimated to be 30.7-percent effective at preventing severe overinflation (at least 25% above manufacturer's recommended cold tire pressure). This may provide further benefits in terms of safety, efficiency, and component wear.

It is difficult to estimate the number of crashes, injuries, or fatalities that are prevented by TPMS. The TPMS-SS provided the necessary data to create an estimate of the reduction in underinflation due to TPMS, but current data sources are not sufficient to create an estimate of the number of crashes caused by underinflation. Future analyses using real world crash data are necessary to evaluate the effect of TPMS on crash involvement.

The increase in fuel economy for vehicles with TPMS estimated in this report is an additional benefit of the technology. Current benefits were estimated at over five hundred million dollars in fleet-wide fuel savings during 2011, and these savings are expected to increase as TPMS saturates the fleet. These savings may also increase if direct systems are replaced with advanced indirect systems that require less maintenance and will not suffer attrition due to battery life span. Conversely, these savings may decrease slightly as average vehicle fuel economy increases.

6.2: Possible Sources of Bias

Model Year: Two factors may have introduced a large bias to the estimates listed above. The first is a model year effect. As shown in Figure 9 below, TPMS is more likely to be found in later model year vehicles.

Fig. 9: Percent of Surveyed Vehicles With TPMS by Model Year

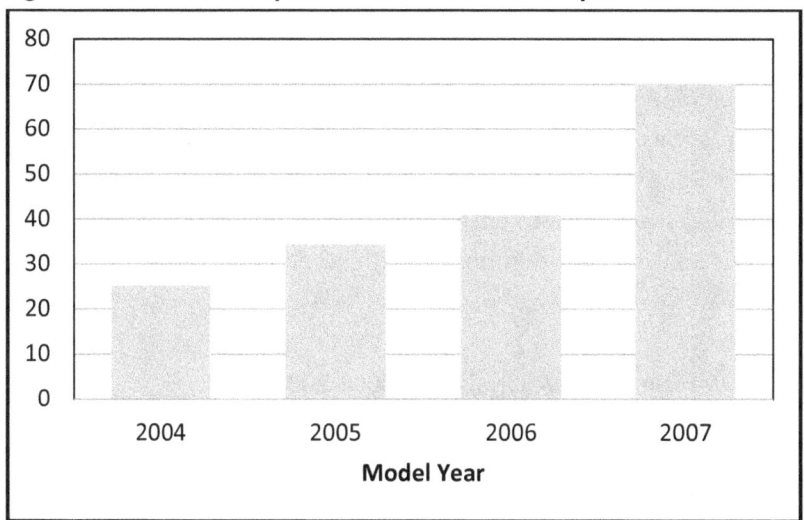

If newer vehicles are more likely to have properly inflated tires for reasons unrelated to TPMS, then this interaction would result in an overestimate of TPMS effectiveness. Fortunately, Figure 3 shows that this is unlikely. Notice that the rate of underinflation in vehicles without TPMS is relatively constant across the included model years. Although there does appear to be a linear model year effect on underinflation rate in the vehicles with TPMS, as discussed earlier this could be due to system attrition or improvements to TPMS technology. If the underinflation rate of vehicles without TPMS is truly constant over time, it would be valid to include the data from model years 2008-2011. However, since this can't be confirmed (there are no vehicles in that model year range without TPMS), these years were excluded resulting in a more conservative estimate of TPMS effectiveness.

Vehicle Model: The second factor that may be introducing bias is a vehicle model effect. It is possible that the vehicle models equipped with TPMS during one of the specific analyzed model years are different from those still without TPMS in the same model year (sportier, more expensive, etc.) and that these differences may cause them to be better maintained. In order to address this possible bias a supplemental analysis was conducted. Although this supplemental analysis is separate and distinct from the primary analysis and its results, it will be described in some detail in the following section because it lends insight into the primary analysis and its validity.

To remove as much of the possible bias introduced by vehicle model the vehicles included in the analysis were restricted to models that met one of the two following criteria:

1) Offered TPMS as an option at some point (MY 2004-2010). Only years with vehicle model observations both with and without TPMS were included.
2) Transitioned from no available TPMS to standard TPMS in consecutive model years. Only the year before and the year after TPMS introduction were included.

The purpose of applying these restrictions is to create a subsample of the survey data that is as free as possible from bias caused by differences between the vehicle models equipped with TPMS and the vehicle models without TPMS. These restrictions will also reduce bias caused by the linear trend observed in TPMS effectiveness across model years. The resulting estimates will be less precise because of a reduced sample size, but they will be less susceptible to both vehicle model and model year sources of bias. The estimates that result from this analysis are risk ratios as in the principal analysis. The results are given in Table 14 below.

Table 14: TPMS Effectiveness – Within Vehicle Model Analysis (MY 2004-2007)

	Any Underinflation	> 10% Underinflation	> 25% Underinflation
Risk ratio	.951	.581*	.39*
Percent Effectiveness (95% CI)	**4.9%** (-4%, 15%)	**42.9%*** (25%, 75%)	**61%*** (4%, 154%)
*=statistically significant at p<.05			

These results are nearly identical to the results of the principal analysis. The confidence intervals are larger because the sample sizes are smaller. The similarity of the within-model results gives strong evidence that there is no significant vehicle model bias in the principal analysis.

6.3: Limitations

As with any survey, it is important to consider the differences between the population of vehicles that had an opportunity to be included in the study and the population to which one wishes to extrapolate. In this case, only vehicles that visited large gas stations between the hours of 8 a.m. and 6 p.m. were exposed to the sampling process. Also, only vehicles of model year 2004-2011 were eligible, and therefore the resulting estimates have no contribution from vehicles more than eight years old.

The methods of analysis were designed to minimize the effects of likely sources of bias. However it is important to note that the effectiveness estimates are created from a very limited range of model year vehicles (2004-2007). Also, effectiveness was averaged across these years despite the appearance of a linear trend in TPMS effectiveness in preventing underinflation over time. The effect of this averaging is that the resulting estimates are likely to be conservative for newer or more recent vehicles, but could be overstated for earlier or older vehicles.

6.4: Future Research

This analysis raised some questions that may benefit from future research. The first is the unknown rate of TPMS attrition. The observed decline in TPMS effectiveness as vehicle age increases may be better understood by determining the rate of functioning to non-functioning TPMS as vehicle age increases. This issue could be addressed by a small follow-up survey, limited to identifying the proportion of vehicles of various model years with functioning TPMS.

The second question is the effect of overinflation on safety. It is reasonable to assume that overinflated tires are less durable from a tread wear standpoint and possibly less safe than tires inflated at the recommended level. Although the rate of severe overinflation in the surveyed population was over 10 percent, there is little published research on identifying and quantifying the risks posed by different levels of overinflation.

The final question is the association between underinflation and crash incidence. With an estimate of the increased probability of a vehicle with severely underinflated tires being involved in a crash, one could estimate the number of crashes prevented by TPMS. This analysis of crash data is currently planned to be the next major step in NHTSA's evaluation of TPMS.

7: Appendix

7.1: Weighting

The TPMS-SS has a complex sampling structure, with several levels of stratification. The unit of observation is the vehicle, and to reach an individual vehicle this study used four separate levels of selection. 1) The primary sampling units (PSU's) are geographic regions within the United States that can represent a city, county, or other region. 2) Within each PSU, several ZIP codes were selected. 3) Within each ZIP code, two eligible gas stations were selected. 4) Within each gas station, several vehicles were selected for observation.

PSU Sampling:

The first stage of weighting at the PSU level has been computed previously and is available in the NASS-CDS data files. These weights are adjusted inverse probabilities of selection and can be combined with the weights from each sampling stage to create the final observation weights.

ZIP Code Sampling:

The ZIP codes are treated as a simple random sample from the frame of ZIP codes within each PSU. This means that although the ZIP codes have different population sizes, each ZIP code within a given PSU has an equal probability of selection. Let $1 \leq i \leq 24$ represent the 24 PSUs included in the study. Let M_i denote the total number of ZIP codes in PSU i, and let $1 \leq j \leq M_i$ be the jth ZIP code of the ith PSU, where m_i denotes the total number of selected ZIP codes in the ith PSU.

Using this notation we can represent the probability that ZIP j in PSU i is included in the TPMS-SS sample using the following formula:

$$P_{ij} = \delta_i \frac{m_i}{M_i}$$

where δ_i denotes the selection probability for the ith PSU (inverse of the PSU weight from the NASS database). Notice that at this stage each sampled ZIP code within a single PSU will have a common probability of selection.

Gas Station Sampling:

Within each selected ZIP code, two gas stations that met the eligibility requirements were selected for data collection. The selection of gas stations within the ZIP codes was done by data collection staff as a convenience sample. Because of the method of selection, the gas stations were considered to be selected with conditional certainty, and each selected station was assigned a weight of 1, or unity. The theory here is that given a certain ZIP code, there will be two gas stations that are immutably the most convenient for the research staff. These two stations are the most convenient regardless of whether or not the ZIP code is selected, and if the ZIP code is selected, it is a certainty in terms of probability that these two and only these two will be selected by the research staff. This method of selection was necessary due to the prohibitive expense of developing an adequate sampling frame, as well as the additional cost of sending data collection teams to non-optimal locations.

The omission of a count of eligible gas stations within each sampled ZIP code means that the resulting final weights can only be used to estimate rates within the sampled population and not the overall size of the population of interest. For example, direct application of the weights will not allow a statement such as 'There are x vehicles in the United States with at least one severely underinflated tire.' Rather, it allows a statement like 'Within the sampled population, p percent of the vehicles had at least one severely underinflated tire.' This estimate may then be combined with existing estimates of the number of vehicles in the United States in order to arrive at the total number of vehicles with underinflated tires in the country.

Vehicle Sampling: The third stage of sampling (vehicle) was conducted on site by the data collection researchers. It was not likely that the researchers would be able to approach every vehicle in the population of interest. This level of selection was largely based on convenience of the data collection researchers, although it was considered pseudo-random and unlikely to bias any estimates. A census of vehicles entering the station was taken during five separate evenly spaced fifteen minute intervals throughout the site observation period.

Let $\widehat{N}_{ijk}$ be the estimated total number of vehicles that enters the kth site of the jth ZIP code of the ith PSU during the eight-hour observation period. Let n_{ijk} be the subset of $\widehat{N}_{ijk}$ that is successfully sampled by the research staff. The vehicle selection probability can then be given by the following formula:

$$P_{ijkl} = \delta_i \frac{m_i}{M_i} \frac{n_{ijk}}{\widehat{N}_{ijk}} \; for \; 1 \leq k \leq M_i \; and \; 1 \leq l \leq \widehat{N}_{ijk}$$

$\widehat{N}_{ijk}$ is estimated as follows: let o_{ijkq} be the total number of vehicles observed during the qth vehicle census period at the kth site of the jth zip code of the ith PSU. Then $\widehat{N}_{ijk}$ is estimated as:

$$\widehat{N}_{ijk} = (6.75/1.25) \sum_{q=1}^{5} o_{ijkq}$$

The sum of observed vehicles during the census periods is multiplied by (6.75/1.25) to reflect the respective amount of time that was spent taking vehicle counts (1.25 hrs) and collecting data (6.75 hrs) during the 8 hour site observation period.

Correction Factor:

As described previously, the ZIP codes within a given PSU were sampled with equal probability even though the ZIP codes have different population sizes. To account for this difference, a correction factor was derived from 2010 Census data, which gives population by ZIP code. Let pop_{ij} be the reported population of the jth ZIP code within the ith PSU. Then the ZIP code size correction factor (ϑ_{ij}) is described as follows:

$$\theta_{ij} = pop_{ij}/(\frac{\sum_{j=1}^{m_i} pop_{ij}}{m_i})$$

Notice that $\sum_{k=1}^{m_i} \theta_{ij} = 1$.

Final Observation Weights:

The final observation weights were applied to the vehicle data at the gas station level (every vehicle observed at a given station had the same weight) and is given by the following equation:

$$w'_{ijkl} = \theta_{ij}/(\delta_i * \frac{m_i}{M_i}\frac{n_{ijk}}{N_{ijk}})$$

This represents the inverse of the probability of selection as well as the ZIP code size correction factor.

7.2: Survey Forms

Paperwork Reduction Act Burden Statement

A federal agency may not conduct or sponsor, and a person is not required to respond to, nor shall a person be subject to a penalty for failure to comply with a collection of information subject to the requirements of the Paperwork Reduction Act unless that collection of information displays a current valid OMB Control Number. The OMB Control Number for this information collection is 2127-0626. Since data will be collected on this form via observation, public reporting for this collection of information is estimated to be approximately 0 minutes per response, including the time for reviewing instructions, completing and reviewing the collection of information. All responses to this collection of information are voluntary. Send comments regarding this burden estimate or any other aspect of this collection of information, including suggestions for reducing this burden to: Information Collection Clearance Officer, National Highway Traffic Safety Administration, 1200 New Jersey Ave, S.E., Washington, DC, 20590. NHTSA Form 1060

U.S. Department of Transportation
National Highway Traffic Safety Administration

DAILY SITE FORM
TALLIES & INSPECTIONS

Form Approved O.M.B. No. 2127-0626
Expiration Date: 06/30/13

National Automotive Sampling System
Tire Pressure Monitoring System-Special Study

1. Primary Sampling Unit Number: ___ ___
2. Site Number: ___ ___
3. Researcher 1: _____
4. Researcher 2: _____
5. Date of Observation: ___ /___ /2010/2011
6. Area: ☐ Urban, ☐ Suburban, ☐ Rural
7. Zip Code ___ ___ ___ ___ ___
8. Spanish Speaker Available: ☐ No ☐ Yes, all day ☐ Yes, partial day
9. Time Period: From ___ To ___

10. VEHICLE BODY TYPES	11. VEHICLE BODY TYPE COUNT TALLIES					12. TALLY TOTAL	13. INSP. TOTAL	14. REFUSAL TOTAL
	PERIOD 1 (08:00 - 08:15)	PERIOD 2 (10:00 - 10:15)	PERIOD 3 (12:00 - 12:15)	PERIOD 4 (02:00 - 02:15)	PERIOD 5 (04:00 - 04:15)			
SMALL AUTOS								
LARGE AUTOS								
UTILITY VEHICLES								
VAN BASED LIGHT TRUCKS								
LIGHT CONV. TRUCKS								
15. TOTALS FOR THE DAY								

16. NOTES

Paperwork Reduction Act Burden Statement

A federal agency may not conduct or sponsor, and a person is not required to respond to, nor shall a person be subject to a penalty for failure to comply with a collection of information subject to the requirements of the Paperwork Reduction Act unless that collection of information displays a current valid OMB Control Number. The OMB Control Number for this information collection is 2127-0626. Since data will be collected on this form via observation, public reporting for this collection of information is estimated to be approximately 0 minutes per response, including the time for reviewing instructions, completing and reviewing the collection of information. All responses to this collection of information are voluntary. Send comments regarding this burden estimate or any other aspect of this collection of information, including suggestions for reducing this burden to: Information Collection Clearance Officer, National Highway Traffic Safety Administration, 1200 New Jersey Ave, S.E., Washington, DC, 20590.   NHTSA Form 1061

U.S. Department of Transportation
National Highway Traffic Safety Administration

DAILY SITE FORM
REFUELING

Form Approved O.M.B. No. 2127-0626
Expiration Date: 06/30/13

National Automotive Sampling System
Tire Pressure Monitoring System-Special Study

1. Primary Sampling Unit Number: ____ ____
2. Site Number: ____ ____
3. Date of Observation: ____/____/2010/2011
4. Station Characteristics: ☐ Cash Only ☐ Pay-at-Pump ☐ Cashier Window ☐ Store ☐ Car Wash ☐ Auto Repair   [All that apply]

Page ____ of ____ Pages

REFUELING

5. Body Type (Check One)	Other Descriptors (e.g. Color)	6. Time In	7. Time Out	8. Activities
Auto: ☐ Small ☐ Large ☐ SUV ☐ Van ☐ PU 1.				☐ Pay and Leave [Only Activity] OR ☐ Store ☐ Auto-Related ☐ Other [All that apply]
Auto: ☐ Small ☐ Large ☐ SUV ☐ Van ☐ PU 2.				☐ Pay and Leave [Only Activity] OR ☐ Store ☐ Auto-Related ☐ Other [All that apply]
Auto: ☐ Small ☐ Large ☐ SUV ☐ Van ☐ PU 3.				☐ Pay and Leave [Only Activity] OR ☐ Store ☐ Auto-Related ☐ Other [All that apply]
Auto: ☐ Small ☐ Large ☐ SUV ☐ Van ☐ PU 4.				☐ Pay and Leave [Only Activity] OR ☐ Store ☐ Auto-Related ☐ Other [All that apply]
Auto: ☐ Small ☐ Large ☐ SUV ☐ Van ☐ PU 5.				☐ Pay and Leave [Only Activity] OR ☐ Store ☐ Auto-Related ☐ Other [All that apply]
Auto: ☐ Small ☐ Large ☐ SUV ☐ Van ☐ PU 6.				☐ Pay and Leave [Only Activity] OR ☐ Store ☐ Auto-Related ☐ Other [All that apply]
Auto: ☐ Small ☐ Large ☐ SUV ☐ Van ☐ PU 7.				☐ Pay and Leave [Only Activity] OR ☐ Store ☐ Auto-Related ☐ Other [All that apply]
Auto: ☐ Small ☐ Large ☐ SUV ☐ Van ☐ PU 8.				☐ Pay and Leave [Only Activity] OR ☐ Store ☐ Auto-Related ☐ Other [All that apply]
Auto: ☐ Small ☐ Large ☐ SUV ☐ Van ☐ PU 9.				☐ Pay and Leave [Only Activity] OR ☐ Store ☐ Auto-Related ☐ Other [All that apply]
Auto: ☐ Small ☐ Large ☐ SUV ☐ Van ☐ PU 10.				☐ Pay and Leave [Only Activity] OR ☐ Store ☐ Auto-Related ☐ Other [All that apply]
Auto: ☐ Small ☐ Large ☐ SUV ☐ Van ☐ PU 11.				☐ Pay and Leave [Only Activity] OR ☐ Store ☐ Auto-Related ☐ Other [All that apply]
Auto: ☐ Small ☐ Large ☐ SUV ☐ Van ☐ PU 12.				☐ Pay and Leave [Only Activity] OR ☐ Store ☐ Auto-Related ☐ Other [All that apply]
Auto: ☐ Small ☐ Large ☐ SUV ☐ Van ☐ PU 13.				☐ Pay and Leave [Only Activity] OR ☐ Store ☐ Auto-Related ☐ Other [All that apply]
Auto: ☐ Small ☐ Large ☐ SUV ☐ Van ☐ PU 14.				☐ Pay and Leave [Only Activity] OR ☐ Store ☐ Auto-Related ☐ Other [All that apply]

Paperwork Reduction Act Burden Statement

A federal agency may not conduct or sponsor, and a person is not required to respond to, nor shall a person be subject to a penalty for failure to comply with a collection of information subject to the requirements of the Paperwork Reduction Act unless that collection of information displays a current valid OMB Control Number. The OMB Control Number for this information collection is 2127-0626. Since data will be collected on this form via observation and inspection, public reporting for this collection of information is estimated to be approximately 0 minutes per response, including the time for reviewing instructions, completing and reviewing the collection of information. All responses to this collection of information are voluntary. Send comments regarding this burden estimate or any other aspect of this collection of information, including suggestions for reducing this burden to: Information Collection Clearance Officer, National Highway Traffic Safety Administration, 1200 New Jersey Ave, S.E., Washington, DC, 20590. NHTSA Form 1062

TIRE INSPECTION FORM

U.S. Department of Transportation
National Highway Traffic Safety Administration

Form Approved O.M.B. No. 2127-0626
Expiration Date: 06/30/13

National Automotive Sampling System
Tire Pressure Monitoring System-Special Study

1. Primary Sampling Unit Number ___ ___
2. Site Number ___ ___
3. Observation Number ___ ___
4. Date of Observation ___ ___ / ___ ___ / 2010/2011
5. Time of Day _____

6. Vehicle Model Year _____
7. Vehicle Make _____
8. Vehicle Model _____
9. Ambient Air Temperature ___ ___
10. Weather: ☐ Clear, ☐ Cloudy, ☐ Fog, ☐ Rain, ☐ Sleet, ☐ Snow

11. TIRE	12. TIRE MANUFACTURER	13. TIRE MODEL	14. TIRE SIZE (eg. P215/70R14)	15. MAXIMUM PRESSURE	16. MEASURED PRESSURE	17. TIRE TEMPERATURE	18. MEASURED MIN. TREAD DEPTH	TIRE
LF				___ ___ psi	___ ___ psi	___ ___ ___ °F	___ ___ /32"	LF
LR				___ ___ psi	___ ___ psi	___ ___ ___ °F	___ ___ /32"	LR
RR				___ ___ psi	___ ___ psi	___ ___ ___ °F	___ ___ /32"	RR
RF				___ ___ psi	___ ___ psi	___ ___ ___ °F	___ ___ /32"	RF

Paperwork Reduction Act Burden Statement

A federal agency may not conduct or sponsor, and a person is not required to respond to, nor shall a person be subject to a penalty for failure to comply with a collection of information subject to the requirements of the Paperwork Reduction Act unless that collection of information displays a current valid OMB Control Number. The OMB Control Number for this information collection is 2127-0626. Since data will be collected on this form via observation and inspection, public reporting for this collection of information is estimated to be approximately 0 minutes per response, including the time for reviewing instructions, completing and reviewing the collection of information. All responses to this collection of information are voluntary. Send comments regarding this burden estimate or any other aspect of this collection of information, including suggestions for reducing this burden to: Information Collection Clearance Officer, National Highway Traffic Safety Administration, 1200 New Jersey Ave, S.E., Washington, DC, 20590. NHTSA Form 1063

U.S. Department of Transportation
National Highway Traffic Safety Administration

VEHICLE INSPECTION FORM

Form Approved O.M.B. No. 2127-0626
Expiration Date: 06/30/13

National Automotive Sampling System
Tire Pressure Monitoring System-Special Study

---

1. Primary Sampling Unit Number ____ ____
2. Site Number ____ ____
3. Observation Number ____ ____
4. Date of Observation ___ ___ / ___ ___ / 2010/2011

VEHICLE IDENTIFICATION

5. Vehicle Model Year _____
6. Vehicle Make _____
7. Vehicle Model _____
8. Vehicle Mileage _____

9. Vehicle Body Type Category
   1) Automobiles: ☐ Small ☐ Large
   2) ☐ Utility Vehicles
   3) ☐ Van Based Light Trucks
   4) ☐ Light Conventional Trucks

10. Vehicle Identification Number (VIN)

    __ __ __ __ __ __ __ __ __ __ __ __ __ __ __ __ __
    1 2 3 4 5 6 7 8 9 10 11 12 13 14 15 16 17

    Left justify; Slash zeros and letter Z (and) —No VIN—Code all zeros —Unknown—Code all nines

TPMS INFORMATION

11. TPMS Display
    1) Display Only   ☐ No, ☐ Yes
    2) Tire Specific Warning Icon: ☐ No, ☐ Yes
    3) Tire Specific PSI: ☐ No, ☐ Yes

        LF Tire ___ ___ psi   LR Tire ___ ___ psi
        RR Tire ___ ___ psi   RF Tire ___ ___ psi

PLACARD/OWNER'S MANUAL INFORMATION

12. GVWR   ___ ___ ___ ___ lbs
13. Manufacturer's Recommended Tire Size*   _____
14. Manufacturer Recommended Cold Tire Pressure (Front)**   ___ ___ psi
15. Manufacturer Recommended Cold Tire Pressure (Rear)**   ___ ___ psi

Paperwork Reduction Act Burden Statement

A federal agency may not conduct or sponsor, and a person is not required to respond to, nor shall a person be subject to a penalty for failure to comply with a collection of information subject to the requirements of the Paperwork Reduction Act unless that collection of information displays a current valid OMB Control Number. The OMB Control Number for this information collection is 2127-0626. Public reporting for this collection of information is estimated to be approximately 10 minutes per response, including the time for reviewing instructions, completing and reviewing the collection of information. All responses to this collection of information are voluntary. Send comments regarding this burden estimate or any other aspect of this collection of information, including suggestions for reducing this burden to: Information Collection Clearance Officer, National Highway Traffic Safety Administration, 1200 New Jersey Ave, S.E., Washington, DC, 20590. NHTSA Form 1064

United States Department of Transportation
National Highway Traffic Safety Administration

INTERVIEW FORM
TIRE PRESSURE

Form Approved O.M.B. No. 2127-0626
Expiration Date: 06/30/13

National Automotive Sampling System
Tire Pressure Monitoring System – Special Study

1. Primary Sampling Unit Number ___ ___
2. Site Number ___ ___
3. Observation Number ___ ___
4. Date of Observation ___ ___ / ___ ___ / 2010/2011
5. Interview in: ○ English ○ Spanish
6. Observations: (○ Interviewed ○ Refused ○ <2004)
   1) Time of Day _____
   2) Ambient Air Temperature ___ ___
   3) Weather: ☐ Clear, ☐ Cloudy, ☐ Fog,
      ☐ Rain, ☐ Sleet, ☐ Snow
   4) Body Type: Auto: ○ Small ○ Large
      ○ SUV ○ Van ○ PU
   5) Sex: ○ Male ○ Female
   6) Age: ○ Young Adult ○ Adult ○ Senior
   7) # in Vehicle: ___ ○ Unknown

[Questions about Vehicle]
7. Who is the owner of this vehicle? (Check One)
   1) ○ Joint with other
   2) ○ Self
   3) ○ Partner/spouse/significant other
   4) ○ Parent or Other family member
   5) ○ Friend or neighbor
   6) ○ Lease
   7) ○ Short-term rental
   8) ○ Car-share
   9) ○ Company/work
   10) ○ Other

8. How long have you had this vehicle?
   Years: ____ Months: ____ Days: ____
   (< 1 month)

9. Was this vehicle new when you obtained it?
   ○ No   ○ Yes

10. Have any of the original tires on this vehicle been replaced? If yes, which ones and when?

10A. Original tires replaced?				10B. If Yes, when?		
Tire	Yes	No	Don't Know	Yrs	Mos	Unk
1) LF	☐	☐	☐			☐
2) LR	☐	☐	☐			☐
3) RR	☐	☐	☐			☐
4) RF	☐	☐	☐			☐
5) Spare	☐	☐	☐			☐
6) Other Specify:	☐	☐	☐			☐

[Questions about tire pressure]
11. Drivers keep their tires at their proper pressure for different reasons. List the reasons that are important to you for keeping tires properly inflated. (Do not read categories, but check all that apply)
    1) ☐ Improved safety
    2) ☐ Improved vehicle performance/handling
    3) ☐ Improved fuel economy
    4) ☐ Longer lasting tires
    5) ☐ Other (specify) _____

12. Where would you, or do you, primarily turn for information on what pressure to set your tires for this vehicle? (Check one)
    1) ○ Intuition/prior knowledge
    2) ○ Owner's manual
    3) ○ Vehicle placard
    4) ○ Tire sidewall labeling
    5) ○ A service technician
    6) ○ OnStar or other automatic system
    7) ○ Relative or friend
    8) ○ Don't know
    9) ○ Other (specify) _____

13. Whose responsibility is it to check the tire pressure? (Check one)
    1) ○ Self
    2) ○ Relative or friend
    3) ○ Service station/dealer
    4) ○ TPMS
    5) ○ OnStar or other automatic system
    6) ○ Owner (other than self, relative or friend)
    7) ○ No one
    8) ○ Other (specify) _____

14. Under what circumstances do you have the tire pressure on this vehicle checked, either by yourself or someone else? (Check all that apply)
    1) ☐ Never (Skip to Q 16—Add Air)
    2) ☐ Before a long trip
    3) ☐ When tires look or feel low
    4) ☐ When tire pressure warning light comes on
    5) ☐ When car is serviced
    6) ☐ When the load being carried is changed
    7) ☐ Tire pressure is checked on a regular basis
    8) ☐ By OnStar or other automatic system
    9) ☐ Don't know
    10) ☐ Other (specify) _____

Paperwork Reduction Act Burden Statement

A federal agency may not conduct or sponsor, and a person is not required to respond to, nor shall a person be subject to a penalty for failure to comply with a collection of information subject to the requirements of the Paperwork Reduction Act unless that collection of information displays a current valid OMB Control Number. The OMB Control Number for this information collection is 2127-0626. Public reporting for this collection of information is estimated to be approximately 10 minutes per response, including the time for reviewing instructions, completing and reviewing the collection of information. All responses to this collection of information are voluntary. Send comments regarding this burden estimate or any other aspect of this collection of information, including suggestions for reducing this burden to: Information Collection Clearance Officer, National Highway Traffic Safety Administration, 1200 New Jersey Ave, S.E., Washington, DC, 20590. NHTSA Form 1065

INTERVIEW FORM
REFUELING

United States Department of Transportation
National Highway Traffic Safety Administration

Form Approved O.M.B. No. 2127-0626
Expiration Date: 06/30/13

National Automotive Sampling System
Tire Pressure Monitoring System – Special Study

1. Primary Sampling Unit Number ___ ___
2. Site Number ___ ___
3. Observation Number ___ ___
4. Date of Observation ___/___/ 2010/2011
5. Interview in: O English   O Spanish
6. Observations: (O Interviewed   O Refused   O <2004)
   1) Time of Day _____
   2) Ambient Air Temperature ___ ___
   3) Weather: ☐ Clear, ☐ Cloudy, ☐ Fog, ☐ Rain, ☐ Sleet, ☐ Snow
   4) Body Type: Auto: O Small O Large   O SUV O Van O PU
   5) Sex: O Male O Female
   6) Age: O Young Adult O Adult O Senior
   7) # in Vehicle: _____ O Unknown

[Questions about Refueling]

7. Did you go out of your way to get to this gas station? If so, how far? _____ (Nearest ¼ mile)
8. Did it take extra time to get to this gas station? If so, how long? _____ (Nearest minute)
9. Before filling up your tank, where was the gas gauge? _____ (Code to nearest 1/8" tank)
10. How many persons total are in this vehicle? _____
11. How many of them are under the age of 16? _____
12. For each of the persons in this vehicle, what is his/her primary reason for traveling?

	Driver	Adults	<16 Yrs.
a. To/From Work			
b. On Work Time			
c. Other			

13. How many gallons of gas did you put in your vehicle? _____ (Code to nearest gallon)
14. After adding gas to your tank, where was the gas gauge? _____ (Code to nearest 1/8" tank)

IF RESPONSE TO QUESTION #14 IS "Full", CONTINUE; IF NOT, SKIP TO QUESTION #16

15. If Full: Do you always fill up your tank?
    O No   O Yes   *If refueling, skip to #16*

16. What is the primary reason you stopped for gas today? (Check one)
    1) O Gas tank low
    2) O Price of the gas
    3) O Fill up on routine basis (e.g., weekly, bi-weekly)
    4) O Top off tank for specific reason (e.g., before long trip)
    5) O Convenient at this time
    6) O To get/do something else (e.g., food, rest stop)
    7) O Other (specify) _____

17. Does this vehicle have a Tire Pressure Monitoring System – also known as a TPMS system?
    1) O No
    2) O Yes
    3) O Don't know

Now I need to ask you some basic information about yourself. [Demographic Information]

18. What is your home zip code? ___ ___ ___ ___ ___
19. How old are you? _____ (Code to nearest yr)
20. What is the highest grade or year of school you completed?
    1) O Less than high school
    2) O High school / GED
    3) O Some college
    4) O College graduate
    5) O Higher degree
    6) O (Vol) Refused

(Continue only for vehicles that have TPMS; Q#17)

21. Would you have time now to answer a few questions on TPMS?
    1) O No (Go to Q 22-Do Later)
    2) O Yes (Go to Supplemental Form)

22. Would you be willing to answer a few questions on TPMS at a later date, using:
    1) O On-line
    2) O Mail-back form
    3) O Phone call back
    4) O Refuse (End)

23. What is your name? _____
24. At what phone number(s) would you like to be called? _____
25. What are good times to call? _____
26. SUP ID: _____

*** Note: Check that INR13-INR15 have been answered ***

United States Department of Transportation
National Highway Traffic Safety Administration

SUPPLEMENTAL FORM

Form Approved O.M.B. No. 2127-0626
Expiration Date: 06/30/13

National Automotive Sampling System
Tire Pressure Monitoring System – Special Study

Paperwork Reduction Act Burden Statement

A federal agency may not conduct or sponsor, and a person is not required to respond to, nor shall a person be subject to a penalty for failure to comply with a collection of information subject to the requirements of the Paperwork Reduction Act unless that collection of information displays a current valid OMB Control Number. The OMB Control Number for this information collection is 2127-0626. Public reporting for this collection of information is estimated to be approximately 10 minutes per response, including the time for reviewing instructions, completing and reviewing the collection of information. All responses to this collection of information are voluntary. Send comments regarding this burden estimate or any other aspect of this collection of information, including suggestions for reducing this burden to: Information Collection Clearance Officer, National Highway Traffic Safety Administration, 1200 New Jersey Ave, S.E., Washington, DC, 20590.   NHTSA Form 1066

Official Use Only
1. Primary Sampling Unit Number ____ ____
2. Site Number ____ ____
3. Observation Number ____ ____
4. Date of Observation ___ ___ / ___ ___ / 2010/2011
5. SUP ID ___ ___ ___ ___ ___ ___ ___ ___ ___

6. Tire Pressure Monitoring Systems (TPMS) can have:
   1) A <u>warning</u> lamp used to indicate low tire pressure
   2) A <u>malfunction</u> lamp used to indicate the TPMS is not working properly.
   3) A <u>combined warning/malfunction</u> lamp used to indicate low tire pressure and/or the system is not working properly.

   Does your TPMS have either a <u>warning</u> lamp or a <u>combined warning/malfunction</u> lamp? *(Check one)*
   O No   O Yes   O Don't Know

   IF RESPONSE TO QUESTION #6 IS "YES", CONTINUE;
   IF NOT, SKIP TO QUESTION #15

[Questions on TPMS Low Pressure Warning Lamp]

7. Do you know where your TPMS <u>warning</u> (combined) lamp is located? If yes, where? *(Check one)*
   1) O No
   2) O Yes, on instrument panel
   3) O Yes, on rearview mirror
   4) O Yes, roof console
   5) O Yes, other (specify) _____

8. Has your TPMS <u>warning</u> (combined) lamp ever illuminated except during engine on/off cycles? If yes, how many times? *(Check one)*
   1) O No
   2) O Yes, ___ ___ *(Approximate number of times)*
   3) O Yes, light is continuously illuminated or comes on regularly
   4) O Yes, don't know how many times.
   5) O Don't know if illuminated

   IF RESPONSE TO QUESTION #8 IS "YES", CONTINUE;
   IF NOT, SKIP TO QUESTION #15

9. When was the last time the <u>warning</u> (combined) lamp illuminated on this vehicle? *(Check one)*
   1) O Within the past month
   2) O 1-3 months ago
   3) O 4 or more months ago
   4) O Continuously/repetitively
   5) O Don't know

10. What actions did you take the last time the TPMS <u>warning</u> (combined) lamp illuminated? *(Check all that apply)*
    1) ☐ Checked tire pressure
    2) ☐ Reset the TPMS
    3) ☐ Took vehicle to the dealer or a service facility
    4) ☐ Added air
    5) ☐ Did nothing *(Skip to #14)*
    6) ☐ Other (specify)_____

11. How long after you first noticed the lamp illuminated, did you take action? *(Check one)*
    1) O During the same trip (e.g., pulled over)
    2) O Later the same day or within several days
    3) O One or more weeks after

12. Did any of the tires need air? If yes, how many? *(Check one)*
    1) O No
    2) O Yes, _____ *(Number of tires)*
    3) O Yes, don't know how many tires.
    4) O Don't know if any of tires needed air.

    IF RESPONSE TO QUESTION #12 IS "YES", CONTINUE;
    IF NOT, SKIP TO QUESTION #14

13. Approximately how much air was needed in each tire? (Estimate on average if multiple tires needed air) *(Check one)*
    1) O Less than 5 PSI
    2) O 5 to 10 PSI
    3) O 10 to 15 PSI
    4) O More than 15 PSI
    5) O Don't know

Additional Questions on the Other Side---Please Turn Over

SUPPLEMENTAL FORM (Continued)

14. Have you or someone else checked the vehicle because the <u>warning</u> (<u>combined</u>) lamp was not working correctly? If yes, what was found to be the reason? *(Check all that apply)*
    1) ☐ No, did not check it
    2) ☐ Yes, needed re-set
    3) ☐ Yes, sensors or other part in the tire not working
    4) ☐ Yes, batteries needed to be changed
    5) ☐ Yes, light bulb needed to be replaced
    6) ☐ Yes, general problem with TPMS system
    7) ☐ Yes, don't know
    8) ☐ Yes, other (specify) _____

---

15. Do you know how to reset (calibrate) your TPMS? If yes, how do you do it? *(Check one)*
    1) ○ No
    2) ○ Yes, use button in vehicle
    3) ○ Yes, follow menu on display
    4) ○ Yes, only dealer/service station can do it
    5) ○ Yes, other (specify) _____

16. When should your TPMS be reset? *(Check all that apply)*
    1) ☐ Never
    2) ☐ When the tire pressure is checked
    3) ☐ When the tire pressure is changed
    4) ☐ When a tire is changed
    5) ☐ When the tires are rotated
    6) ☐ Don't know
    7) ☐ Other (specify) _____

17. How easy or difficult is it to reset your TPMS? *(Check one)*
    1) ○ Very easy
    2) ○ Somewhat easy
    3) ○ Somewhat difficult
    4) ○ Very difficult
    5) ○ Don't know

18. To what extent do you rely on your TPMS to tell you when your tires need air? *(Check one)*
    1) ○ Rely fully on the TPMS
    2) ○ Rely on TPMS, but also use other methods
    3) ○ Don't rely on TPMS, only use other methods

19. Does your TPMS have a <u>malfunction</u> lamp? *(Check one)*
    ○ No ○ Yes ○ Don't Know

IF RESPONSE TO QUESTION #19 IS "YES", CONTINUE; IF NOT, GO TO THE END

[Questions on TPMS Malfunction Warning Lamp]

20. Do you know where your TPMS <u>malfunction</u> lamp is located? If yes, where? *(Check one)*
    1) ○ No
    2) ○ Yes, on instrument panel
    3) ○ Yes, on rearview mirror
    4) ○ Yes, roof console
    5) ○ Yes, other (specify) _____

IF RESPONSE TO QUESTION #20 IS "YES", CONTINUE; IF NOT, GO TO THE END

21. Has your TPMS <u>malfunction</u> lamp ever illuminated, except during engine on/off cycles? If yes, how many times? *(Check one)*
    1) ○ No
    2) ○ Yes, _____ (Approximate number of times)
    3) ○ Yes, light is continuously illuminated or comes on regularly
    4) ○ Yes, don't know how many times.
    5) ○ Don't know if illuminated

IF RESPONSE TO QUESTION #21 IS "YES", CONTINUE; IF NOT, GO TO THE END

22. When was the last time the <u>malfunction</u> lamp illuminated on this vehicle? *(Check one)*
    1) ○ Within the past month
    2) ○ 1-2 months ago
    3) ○ 3-4 months ago
    4) ○ More than 4 months ago
    5) ○ Continuously/repetitively
    6) ○ Don't know

23. What actions did you take the last time the TPMS <u>malfunction</u> lamp illuminated? *(Check all that apply)*
    1) ☐ Did nothing-it often illuminates
    2) ☐ Did nothing-other reasons
    3) ☐ Reset the TPMS
    4) ☐ Took vehicle to the dealer or a service facility
    5) ☐ Self or others worked on vehicle
    6) ☐ Other (specify) _____

24. Have you or someone else checked the vehicle because the <u>malfunction</u> lamp was not working correctly? If yes, what was found to be the reason? *(Check all that apply)*
    1) ☐ No, did not check it
    2) ☐ Yes, needed re-set
    3) ☐ Yes, sensors or other part in the tire not working
    4) ☐ Yes, batteries needed to be changed
    5) ☐ Yes, light bulb needed to be replaced
    6) ☐ Yes, general problem with TPMS system
    7) ☐ Yes, don't know
    8) ☐ Yes, other (specify) _____

THE END

THANK YOU FOR YOUR PARTICIPATION

DOT HS 811 681
November 2012

U.S. Department
of Transportation
**National Highway
Traffic Safety
Administration**

8915-110212-v4

SUPPLEMENTAL FORM (Continued)

14. Have you or someone else checked the vehicle because the <u>warning</u> (combined) lamp was not working correctly? If yes, what was found to be the reason? *(Check all that apply)*
    1) ☐ No, did not check it
    2) ☐ Yes, needed re-set
    3) ☐ Yes, sensors or other part in the tire not working
    4) ☐ Yes, batteries needed to be changed
    5) ☐ Yes, light bulb needed to be replaced
    6) ☐ Yes, general problem with TPMS system
    7) ☐ Yes, don't know
    8) ☐ Yes, other (specify) _____

15. Do you know how to reset (calibrate) your TPMS? If yes, how do you do it? *(Check one)*
    1) ○ No
    2) ○ Yes, use button in vehicle
    3) ○ Yes, follow menu on display
    4) ○ Yes, only dealer/service station can do it
    5) ○ Yes, other (specify) _____

16. When should your TPMS be reset? *(Check all that apply)*
    1) ☐ Never
    2) ☐ When the tire pressure is checked
    3) ☐ When the tire pressure is changed
    4) ☐ When a tire is changed
    5) ☐ When the tires are rotated
    6) ☐ Don't know
    7) ☐ Other (specify) _____

17. How easy or difficult is it to reset your TPMS? *(Check one)*
    1) ○ Very easy
    2) ○ Somewhat easy
    3) ○ Somewhat difficult
    4) ○ Very difficult
    5) ○ Don't know

18. To what extent do you rely on your TPMS to tell you when your tires need air? *(Check one)*
    1) ○ Rely fully on the TPMS
    2) ○ Rely on TPMS, but also use other methods
    3) ○ Don't rely on TPMS, only use other methods

19. Does your TPMS have a <u>malfunction</u> lamp? *(Check one)*
    ○ No ○ Yes ○ Don't Know

IF RESPONSE TO QUESTION #19 IS "YES", CONTINUE; IF NOT, GO TO THE END

[Questions on TPMS Malfunction Warning Lamp]

20. Do you know where your TPMS <u>malfunction</u> lamp is located? If yes, where? *(Check one)*
    1) ○ No
    2) ○ Yes, on instrument panel
    3) ○ Yes, on rearview mirror
    4) ○ Yes, roof console
    5) ○ Yes, other (specify) _____

IF RESPONSE TO QUESTION #20 IS "YES", CONTINUE; IF NOT, GO TO THE END

21. Has your TPMS <u>malfunction</u> lamp ever illuminated, except during engine on/off cycles? If yes, how many times? *(Check one)*
    1) ○ No
    2) ○ Yes, ____ (Approximate number of times)
    3) ○ Yes, light is continuously illuminated or comes on regularly
    4) ○ Yes, don't know how many times.
    5) ○ Don't know if illuminated

IF RESPONSE TO QUESTION #21 IS "YES", CONTINUE; IF NOT, GO TO THE END

22. When was the last time the <u>malfunction</u> lamp illuminated on this vehicle? *(Check one)*
    1) ○ Within the past month
    2) ○ 1-2 months ago
    3) ○ 3-4 months ago
    4) ○ More than 4 months ago
    5) ○ Continuously/repetitively
    6) ○ Don't know

23. What actions did you take the last time the TPMS <u>malfunction</u> lamp illuminated? *(Check all that apply)*
    1) ☐ Did nothing-it often illuminates
    2) ☐ Did nothing-other reasons
    3) ☐ Reset the TPMS
    4) ☐ Took vehicle to the dealer or a service facility
    5) ☐ Self or others worked on vehicle
    6) ☐ Other (specify) _____

24. Have you or someone else checked the vehicle because the <u>malfunction</u> lamp was not working correctly? If yes, what was found to be the reason? *(Check all that apply)*
    1) ☐ No, did not check it
    2) ☐ Yes, needed re-set
    3) ☐ Yes, sensors or other part in the tire not working
    4) ☐ Yes, batteries needed to be changed
    5) ☐ Yes, light bulb needed to be replaced
    6) ☐ Yes, general problem with TPMS system
    7) ☐ Yes, don't know
    8) ☐ Yes, other (specify) _____

THE END

THANK YOU FOR YOUR PARTICIPATION

DOT HS 811 681
November 2012

U.S. Department
of Transportation

**National Highway
Traffic Safety
Administration**

8915-110212-v4

www.ingramcontent.com/pod-product-compliance
Lightning Source LLC
Chambersburg PA
CBHW081858170526
45167CB00007B/3060